SpringerBriefs in Plant Science

SpringerBriefs present concise summaries of cutting-edge research and practical applications across a wide spectrum of fields. Featuring compact volumes of 50 to 125 pages, the series covers a range of content from professional to academic. Typical topics might include:

- A timely report of state-of-the art analytical techniques
- A bridge between new research results, as published in journal articles, and a contextual literature review
- A snapshot of a hot or emerging topic
- An in-depth case study or clinical example
- A presentation of core concepts that students must understand in order to make independent contributions

SpringerBriefs in Plant Sciences showcase emerging theory, original research, review material and practical application in plant genetics and genomics, agronomy, forestry, plant breeding and biotechnology, botany, and related fields, from a global author community. Briefs are characterized by fast, global electronic dissemination, standard publishing contracts, standardized manuscript preparation and formatting guidelines, and expedited production schedules.

Pankaj Kumar

Totipotency in Plants

Molecular Signaling and Gene Network Dynamics

Springer

Pankaj Kumar
Department of Biotechnology
Dr. Yashwant Singh Parmar University of Horticulture and Forestry
Solan, Himachal Pradesh, India

ISSN 2192-1229 ISSN 2192-1210 (electronic)
SpringerBriefs in Plant Science
ISBN 978-981-92-1630-7 ISBN 978-981-92-1631-4 (eBook)
https://doi.org/10.1007/978-981-92-1631-4

This Springer imprint is published by the registered company Springer Nature Singapore Pte Ltd.
The registered company address is: 152 Beach Road, #21-01/04 Gateway East, Singapore 189721, Singapore

Preface

Plant cell totipotency is the intrinsic ability of a single somatic cell to regenerate a complete organism, underpins all modern plant tissue culture, genetic transformation, genome editing, and synthetic biology. This textbook, *Totipotency in Plants: Molecular Signaling and Gene Network Dynamics*, provides a comprehensive and unified framework for understanding the molecular basis of plant cell totipotency, with a specific focus on the gene regulatory networks that govern in vitro regeneration. As the demand for efficient and reproducible plant regeneration systems continues to grow across agriculture, biotechnology, and conservation, a foundational understanding of how plant cells transition to a totipotent state is both timely and essential. Despite major advancements in tissue culture technologies, many economically important species remain recalcitrant due to incomplete knowledge of the underlying genetic controls. Based on the latest research, the book integrates developmental regulatory (DR) genes, hormonal crosstalk, epigenetic reprogramming, and systems biology approaches. It begins with the core transcription factors—*WUSCHEL, BABY BOOM (BBM), LEAFY COTYLEDON (LEC), WOX, GROWTH-REGULATING FACTOR (GRF)*, and *SOMATIC EMBRYOGENESIS RECEPTOR KINASE (SERK)*—and systematically explores their interactions within complex hormonal, chromatin, and network-level controls. Emphasis is placed on experimentally validated functions across model and crop species, including *Arabidopsis*, rice, and wheat, and on practical strategies to modulate these networks through culture conditions, genetic engineering, and synthetic biology.

The chapters progress logically from foundational concepts (Chap. 1) to transcriptional switches (Chap. 2), hormonal modules (Chap. 3), epigenetic mechanisms (Chap. 4), gene networks (Chap. 5), evolutionary perspectives (Chap. 6), and synthetic engineering directions (Chap. 7). Written with clarity and conceptual depth, this textbook is intended for advanced undergraduates, graduate students, and early-career researchers in plant molecular biology, biotechnology, and related fields. It also serves as a reference for tissue culture professionals and crop engineers aiming to implement molecular strategies to improve in vitro regeneration efficiency. By treating totipotency as an orchestrated molecular phenomenon, this textbook empowers readers to translate foundational knowledge into practical applications for plant transformation, genome editing, and synthetic biology.

Solan, Himachal Pradesh, India Pankaj Kumar

Acknowledgments

I express my sincere gratitude to my master's and Ph.D. supervisor, **Professor Dinesh Kumar Srivastava**, whose pioneering work *"Plant Tissue Culture and Genetic Transformation"* has been a profound source of scientific inspiration in shaping my understanding of plant tissue culture mechanisms. Although I first encountered the concept of totipotency during my school days and explored it further during my graduate studies, it was under his guidance that I developed a deeper scientific perspective. His mentorship was instrumental in helping me understand the critical determinants of successful plant regeneration, as well as in designing and executing meaningful research experiments.

The process of growing plants in a glass jar from a tiny explant has always fascinated me. While it appears simple, it raises fundamental questions about how such regeneration occurs. This curiosity evolved into a research pursuit that continues to this day. A thorough examination of the scientific literature reveals that many questions in this domain remain unanswered. This realization has been both motivating and humbling, and it inspired me to write this textbook in plant and agricultural biotechnology, with the hope that it will serve as a seed for future research and innovation.

I extend my heartfelt thanks to my postdoctoral mentors, **Professor (Dr.) Asis Datta**, *Padma Shri*, *Padma Bhushan*, and *Shanti Swaroop Bhatnagar Awardee*, Ex-Vice Chancellor, *Jawaharlal Nehru University*, and Founder Director, *National Institute of Plant Genome Research (NIPGR)*, New Delhi, and **Dr. Sanjay Kumar**, *Chairman, ASRB*, and former *Director, CSIR-IHBT, Palampur*, for their distinguished mentorship, encouragement, and for enriching my practical understanding of plant molecular biology. I am also grateful to **Dr. Ashish R. Warghat (Principal Scientist)** for providing me with the opportunity to engage in collaborative research at CSIR-IHBT, which significantly enhanced my expertise in plant molecular biology and biotechnology.

I sincerely acknowledge the Director of Research and the Vice-Chancellor of **Dr. Y.S. Parmar University of Horticulture and Forestry, Nauni, Solan**, for providing a stimulating and supportive academic environment that facilitated the successful completion of this work.

I gratefully acknowledge the **Science and Engineering Research Board (SERB), Government of India**, now reconstituted as the **Anusandhan National Research Foundation (ANRF)**, for supporting my research as a Young Scientist (NPDF) and Prime Minister Early Career Research Grant Scheme (ECRG). I also acknowledge the support received under the **RKVY-RAFTAAR** program for research on walnut micropropagation.

A special note of appreciation goes to my research scholars, **Deepika**, **E. Manikandan**, and **Shagun Sharma**, whom I fondly regard as budding scientists. Their valuable discussions, assistance in drafting chapters, and contributions to research on important crops such as walnut and *Fritillaria roylei* (Jangli Lahsun) have been immensely helpful.

Finally, I express my deepest gratitude to my parents, my wife **Dr. Shivanti Negi Sharma**, and my daughter **Jiva Sharma** for their unwavering love, patience, and moral support. Their encouragement has been the true foundation of my scientific journey and the completion of this book.

Competing Interests The author has no competing interests to declare that are relevant to the content of this manuscript.

Contents

About the Author

Pankaj Kumar is currently serving as an Assistant Professor in the Department of Biotechnology at Dr. Yashwant Singh Parmar University of Horticulture and Forestry, Solan, Himachal Pradesh, India. With over a decade of research experience in plant molecular biology and biotechnology, he has made notable contributions to Plant science. He did his postdoctoral research at the National Institute of Plant Genome Research (NIPGR), New Delhi, and CSIR-Institute of Himalayan Bioresource Technology (IHBT), Palampur, HP, India. He is a prolific author with over 100 research, review papers, book chapters, and several notable books, on topics including plant stress memory, medicinal plant agriculture, climate resilience and molecular adaptation in alpine medicinal plants, agricultural biotechnology, and ornamental horticulture. His innovative contributions have also led to the publication of five Indian patents. He has received numerous prestigious awards, including the Anusandhan National Research Foundation - Prime Minister Early Career Research Grant (2026), the Young Scientists Fellowship Award (2021) from ICMR, and the Scientist's Pool Scheme Award (2020) from CSIR. His academic leadership extends to active editorial and peer-review roles, and he continues to mentor students and researchers in interdisciplinary plant biotechnology.

Chapter 1
Plant Cell Totipotency

Abstract Plant totipotency is the capacity of a differentiated somatic cell to regenerate into a complete organism that mainly constitutes the biological foundation of plant tissue culture and modern plant biotechnology. This chapter combines current understanding of the cellular, molecular, and epigenetic mechanisms that trigger totipotent reprogramming in in vitro conditions. Totipotency is initiated through stress and wound-induced cellular plasticity, followed by hormone-mediated dedifferentiation and redifferentiation processes governed by auxin-cytokinin interactions. Central to this transition is the activation of developmental regulatory genes, including *LEAFY COTYLEDON* (*LEC1* and *LEC2*), *WUSCHEL* (*WUS*), *ABA INSENSITIVE3* (*ABI3*), *AGAMOUS-LIKE 15* (*AGL15*), *BABY BOOM* (*BBM*), and *FUSCA3* (*FUS3*), which function within interconnected transcriptional networks to provide embryogenic competence. Additionally, epigenetic changes, particularly chromatin remodeling mediated by histone acetylation and modulation of Polycomb Repressive Complex 2 (*PRC2*), facilitates transcriptional reactivation of embryogenic programs. Beyond its developmental significance, totipotency underpins crop genetic transformation, germplasm conservation, genome editing, and synthetic biology applications. By integrating genetic, hormonal, and chromatin-level regulation, this chapter highlights totipotency as a dynamic and programmable developmental state essential to both fundamental plant biology and translational agricultural innovation.

Keywords Developmental regulatory genes · Epigenetics · Molecular mechanisms · Plant tissue culture · Somatic embryogenesis · Totipotency

P. Kumar, *Totipotency in Plants*, SpringerBriefs in Plant Science,
https://doi.org/10.1007/978-981-92-1631-4_1

1.1 Introduction

Plant tissue culture has emerged as a powerful and widely utilized tool for clonal propagation in modern agriculture and forestry, as well as in transgenic engineering. This technology facilitates the rapid mass multiplication of elite genotypes and the preservation of rare germplasm, offering a sophisticated alternative to traditional breeding methods (Morinaka et al. 2023; Bao et al. 2024). Fundamental to these biotechnological advancements is the phenomenon of plant regeneration, a process by which whole plants can be regenerated from mature differentiated tissues, organs, unorganized mass of callus, or even a single cell. Plant regeneration is referred to as physiological repair or replacement and renewal of tissue in plants. The ability of plants to regenerate is based on their totipotency or pluripotency of stem cells, which reflects the high degree of cellular differentiation and developmental flexibility (Long et al. 2022). The word totipotency is derived from the Latin *totipotens*, which means all-powerful. In terms of biology, it describes the astonishing capacity of a single cell to divide and generate every differentiated cell into tissues and organs and ultimately regenerate an entire new individual. Although animal cells largely lose this ability in early embryonic developmental stages, differentiated plant cells have the capacity to retain prior developmental plasticity, i.e., dedifferentiation, and can regain totipotency and pluripotency. In plants, embryogenic competence is not restricted to the zygote. While zygotic embryogenesis normally requires fertilization, the formation of apomictic embryos in certain species provides compelling evidence that embryos can develop in vivo in the absence of sexual reproduction (Barcaccia et al. 2020). Furthermore, in epiphyllous species, somatic cells of vegetative organs are capable of spontaneously giving rise to ectopic embryos under natural conditions. These events require coordinated cellular reprogramming and epigenetic changes that helps in restoring pluripotency and redirecting plant cell fate toward embryo development. Over the past several decades, considerable advances have been achieved in elucidating the mechanisms by which differentiated somatic cells reacquire totipotency. Current research has provided a foundational understanding of the genetic and epigenetic regulatory frameworks that govern the reprogramming of somatic cells into a totipotent state.

1.2 History of Plant Totipotency

Historically, the earliest insights into the mechanisms by which plant cells regulate totipotency were derived from empirical studies employing in vitro culture techniques. In 1902, Gottlieb Haberlandt introduced the concept of plant cell totipotency, building upon the cell theory proposed by Schleiden (1838) and Schwann (1839). He postulated that complete plants could be regenerated from isolated somatic cells (Su et al. 2021). However, experimental validation of this hypothesis was not achieved until 1958, when Steward and colleagues demonstrated that whole plants could be regenerated from segments of differentiated secondary phloem tissue of carrot (*Daucus carota*), thereby providing definitive evidence for the totipotent capacity of somatic plant cells (Su et al. 2021; Chong 2025). The earliest successful reports of plant tissue culture emerged in 1902, when Gottlieb Haberlandt demonstrated the development and maintenance of mesophyll cells exhibiting totipotent characteristics under in vitro conditions, and the term totipotency was coined by T.H. Morgan in 1901. Owing to his foundational contributions, the German botanist Gottlieb Haberlandt is widely regarded as the father of plant tissue culture. Haberlandt further advanced the field by cultivating palisade parenchyma tissues using Knop's salt solution. Shortly thereafter, Hanning (1904) successfully excised and cultured mature plant embryos in vitro on a mineral salt and sugar medium, marking a significant milestone in the development of embryo culture techniques. Since this pioneering work, plant tissue culture techniques have undergone continuous refinement, with numerous studies highlighting their applications in crop improvement and conservation biology through genetic transformation, plant breeding programs, and biopharmaceutical production, thereby forming the conceptual backbone of modern plant biotechnology (Malabadi et al. 2025; Wojcik 2025). The large-scale application of plant tissue culture emerged in the 1950s, particularly driven by its adoption within the orchid industry for commercial propagation. Subsequent breakthroughs continued to expand the scope of tissue culture technologies. Notably, in 1939 White documented the first reproducible differentiation of shoot buds within tobacco tissue cultures. Further, Vasil and Hildebrandt successfully achieved the regeneration of complete plants from isolated single tobacco cells in 1965 (Bhojwani and Dantu 2013). In 1972, Carlson and colleagues achieved the first successful somatic hybridization by fusing protoplasts of *Nicotiana glauca* and *N. langsdorffii*, thereby creating the first somatic hybrid plant and opening new ways for genetic manipulation and crop improvement. This inherent cellular plasticity has facilitated the development of efficient plant tissue culture protocols, thereby enabling the breeding of crop varieties with enhanced stress tolerance, increased productivity, and improved adaptability.

While traditional methodologies for inducing regeneration focused primarily on the exogenous manipulation of culture media, the last decades have shifted the paradigm toward molecular genetics. A deeper understanding of the molecular mechanisms underlying plant regeneration and reprogramming would allow precise regulation of this process, promoting the development of high-yielding and stress-resilient cultivars while significantly accelerating breeding programs across a wide range of plant species (Ibanez et al. 2020). Significant progress has been made in identifying specific genes responsible for shoot and root differentiation, particularly within the model organism *Arabidopsis thaliana* (Bhojwani and Dantu 2013).

1.3 Totipotency in Plant Tissue Culture Systems

In plant tissue culture, plant cells exhibit a significant capacity for totipotency upon wounding, that is, the ability of a single differentiated cell to dedifferentiate and then redifferentiate to form an entire plant. This process is crucial for plant development and adaptation and involves complex cellular and molecular reprogramming events.

1.3.1 Cellular Basis of Totipotency: Plasticity and Reprogramming

Plant totipotency in vitro is fundamentally rooted in cellular plasticity, i.e., the capacity of differentiated somatic cells to revert to a less specialized state and subsequently acquire a new developmental fate. Cellular reprogramming typically begins with dedifferentiation, during which mature cells lose specialized characteristics and re-enter the cell cycle. This is followed by redifferentiation, where cells adopt embryogenic or organogenic identities under defined inductive conditions (Su et al. 2021). Somatic embryogenesis is a dominant method for studying totipotency and plant regeneration. It refers to the process by which somatic cells are reprogrammed to form embryos and represents one of the most remarkable manifestations of cellular plasticity in plants (Chong 2025). Somatic embryos can originate from various differentiated cell types, including ovular, embryonic, and leaf cells, in response to exogenous and endogenous stimuli, as shown in Fig. 1.1. In plant tissue

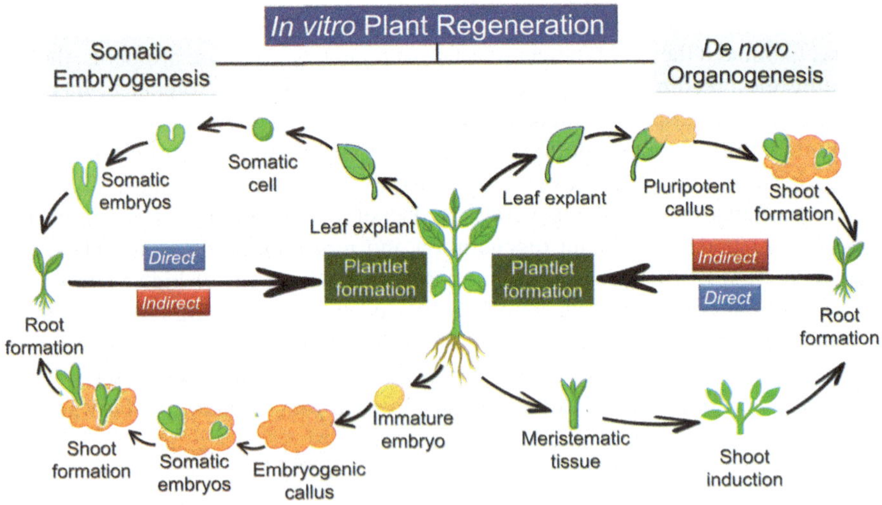

Fig. 1.1 In vitro plant regeneration pathways illustrating cellular totipotency through somatic embryogenesis and de novo organogenesis

culture, somatic embryogenesis is an extensively utilized method for clonal propagation and transgenic engineering in agriculture and forestry. Additionally, it elucidates the cellular and molecular mechanisms of normal zygotic embryogenesis, with its success governed by a dynamic interplay between plant hormone balance, reactive oxygen species localization, and the underlying genetic background (Su et al. 2021).

1.3.2 Molecular Mechanisms Underlying Totipotent Reprogramming

In tissue culture systems, mechanical wounding and culture-associated stress signals initiate reprogramming. These stimuli activate endogenous signaling cascades, particularly those mediated by auxin and cytokinin (Zhu et al. 2023). Plant hormones, particularly auxin and cytokinin, are key regulators used to induce somatic embryogenesis. The synthetic auxin 2,4-dichlorophenoxyacetic acid (2,4-D) is a widely used hormonal stimulus due to its strong auxin-like activity. Endogenous auxin levels are critical for inducing somatic embryogenesis, with a transient increase in indole-3-acetic acid (IAA) often observed. The removal of 2,4-D from the culture medium activates auxin biosynthetic enzymes, leading to increased endogenous IAA levels and promoting somatic embryo formation. Auxin acts as a common inducer for reprogramming somatic cells into totipotent cells (Su et al. 2021). Other hormones such as abscisic acid (ABA) and ethylene can also induce somatic embryogenesis in certain cases; however, ethylene can sometimes interfere with the process. Exogenous supplementation of these hormones re-establishes proliferative competence, induces embryogenic callus formation, and organizes the patterned cell division required for shoot, root, or somatic embryo formation (Méndez-Hernández et al. 2019). Additionally, various stress conditions, such as osmotic stress, heavy metal stress, temperature stress, UV radiation, and wounding, can also trigger or promote somatic embryogenesis (Spinoso-Castillo and Bello-Bello 2022). Importantly, totipotent competence is not constitutively expressed in all plant cells; rather, it represents a reacquired physiological state that depends on specific molecular and environmental cues.

1.4 Activation of Developmental Regulatory Genes

The acquisition of totipotency in plant cells is tightly controlled by a set of developmental regulatory genes that function as master transcriptional switches during cellular reprogramming. Those transcription factors that are crucial for reprogramming somatic cells into a totipotent state include *LEAFY COTYLEDON1* (*LEC1*), *LEAFY COTYLEDON2* (*LEC2*), *FUSCA3* (*FUS3*), *ABA INSENSITIVE3* (*ABI3*), *BABY*

BOOM (*BBM*), *AGAMOUS-LIKE 15* (*AGL15*), and *WUSCHEL* (*WUS*) (Yuan et al. 2024). These regulators orchestrate the transition from a differentiated somatic state to an embryogenic or organogenic state by activating embryo-specific gene expression programs and modulating hormone biosynthesis and signaling pathways. These transcription factors work within an interconnected regulatory network rather than functioning independently. For example, *BBM* acts upstream in the hierarchy and transcriptionally activates *LEC1*, *LEC2*, *FUS3*, *ABI3*, and *AGL15*, thereby promoting embryogenic competence (Xu et al. 2024). *LEC1* and *LEC2* are central integrators of embryonic identity and, when ectopically expressed or derepressed, can initiate somatic embryogenesis in vegetative tissues (Dai et al. 2026). Mechanistically, *LEC2* stimulates auxin biosynthesis by directly activating *YUCCA* (*YUC2*, *YUC4*, *YUC10*) and *TRYPTOPHAN AMINOTRANSFERASE OF ARABIDOPSIS1* (*TAA1*), thereby elevating endogenous auxin levels and reducing dependence on exogenous auxin supplementation (Cao et al. 2019). Through these coordinated transcriptional and hormonal interactions, developmental regulatory genes establish the molecular framework required for totipotent cell specification.

1.5 Epigenetic Reprogramming

In addition to transcriptional activation of developmental regulatory genes, epigenetic remodeling is a fundamental requirement for the establishment of totipotency. Epigenetic mechanisms regulate chromatin accessibility and determine whether embryogenic regulators can be transcriptionally activated (Wang et al. 2020). Repressive chromatin states must be relaxed to permit the expression of embryo-specific genes during somatic embryogenesis. Polycomb Repressive Complex 2 (*PRC2*), an epigenetic regulator, represses the expression of embryo maturation genes and the ectopic accumulation of embryonic storage molecules. Loss of *PRC2* function can promote the reprogramming of somatic cells toward a totipotent state, thereby facilitating the initiation of somatic embryogenesis. Similarly, mutations affecting *PRC2* subunits often result in the formation of unorganized cell masses and the ectopic development of somatic embryos, reflecting the critical role of *PRC2* in maintaining transcriptional repression of embryogenic programs (Su et al. 2021). Histone acetylation is another epigenetic modification observed during cellular differentiation and embryogenic transition. Increased levels of acetylated lysines in histone H3 are correlated with chromatin decondensation and transcriptional activation of genes in cells competent for fate switching (Morończyk et al. 2022). Chemical inhibitors of histone deacetylases, such as trichostatin A (TSA), can induce somatic embryogenesis, highlighting the critical role of histone acetylation in the transition to a totipotent state (Su et al. 2021). Together, these findings indicate that totipotency is achieved through a coordinated interplay between transcriptional regulators and dynamic chromatin modifications.

1.6 Translational and Applied Significance

Plant totipotency is not only a fundamental biological property but also a cornerstone of modern plant biotechnology. The capacity of somatic cells to regenerate into complete plants provides the experimental and technological foundation for genetic transformation, germplasm conservation, genome editing, and synthetic developmental engineering (Wang et al. 2025). A mechanistic understanding of the molecular basis of totipotency, mainly the regulation of developmental regulatory genes, hormonal signaling networks, and epigenetic remodeling, has significantly expanded its translational potential in agriculture and plant science.

1.6.1 Crop Improvement

Totipotency is indispensable for crop genetic transformation and genome engineering. Most transformation systems, including *Agrobacterium*-mediated transformation and biolistic gene delivery, rely on the ability of transformed somatic cells to regenerate into whole fertile plants. Without efficient regeneration, stable integration of transgenes or precise genome edits cannot be translated into improved cultivars (Wu et al. 2024). The advent of CRISPR/Cas-based genome editing has further emphasized the importance of regeneration competence. Successful editing of elite or recalcitrant genotypes requires robust in vitro regeneration systems, which are often genotype-dependent (Kowalik et al. 2026). Overexpression of developmental regulators such as *BBM* and *WUS* has been shown to enhance transformation efficiency in cereals like maize by promoting somatic embryogenesis and bypassing regeneration bottlenecks (Jiang et al. 2024). Thus, molecular manipulation of totipotency regulators provides a strategy to overcome recalcitrance and increase transformation efficiency in economically important crops.

1.6.2 Germplasm Conservation

Totipotency-based regeneration systems are essential for the conservation of plant genetic resources, particularly for rare, endangered, and vegetatively propagated species. Micropropagation enables rapid clonal multiplication under sterile conditions, ensuring genetic fidelity and pathogen-free propagation. This is particularly valuable for woody perennials and horticultural crops with limited seed viability or a long juvenile period (Bettoni et al. 2024). Cryopreservation strategies also depend on the totipotent capacity of preserved tissues. Shoot tips, embryogenic callus, or somatic embryos can be cryostored in liquid nitrogen and later regenerated into complete plants, safeguarding biodiversity and elite germplasm lines (Sirangi

2025). Molecular understanding of stress-induced reprogramming and epigenetic stability during regeneration further improves post-cryopreservation recovery and phenotypic stability.

1.6.3 Synthetic Biology and Regeneration Engineering

The expanding field of plant synthetic biology increasingly relies on precise control over developmental programs, including the induction and modulation of totipotency. By engineering transcriptional circuits and synthetic promoters that regulate key developmental genes such as *LEC*, *BBM*, and *WUS*, researchers can design inducible regeneration systems tailored to specific species or genotypes (Xu et al. 2024). Recent advances in synthetic gene regulatory networks and epigenome editing tools allow targeted activation or repression of totipotency-associated loci. For example, programmable transcriptional activators based on CRISPR/dCas systems can be used to modulate endogenous developmental regulators without stable transgene integration (Pan et al. 2021). Such approaches open possibilities for controlled, reversible induction of embryogenic competence and for constructing minimal totipotency modules that function across species boundaries. Future research integrating transcriptomics, epigenomics, and synthetic regulatory engineering will further refine the ability to modulate totipotent states, positioning plant regeneration as a programmable platform for sustainable agriculture and next-generation crop biotechnology.

1.7 Conclusion

Plant totipotency represents a highly coordinated developmental phenomenon arising from the interplay between cellular plasticity, hormonal signaling, transcriptional control, and epigenetic reprogramming. Instead of constituting an intrinsic and constitutively expressed property of all plant cells, totipotent competence is re-established only under specific physiological and environmental conditions. In vitro systems have provided critical insights into this process, demonstrating how stress cues such as wounding and exogenous hormones initiate dedifferentiation and activate master developmental regulators. Moreover, epigenetic remodeling enables the transcriptional activation of embryogenic pathways. These molecular insights have substantial translational implications. Efficient regeneration is essential for transgenic engineering, CRISPR-based genome editing, germplasm conservation, and synthetic developmental programming. Integrating molecular genetics, epigenomics, and synthetic regulatory engineering may enable customized regeneration strategies for elite cultivars and non-model species, accelerating crop improvement and sustainable agriculture. Consequently, understanding the molecular mechanisms governing totipotency not only advances fundamental developmental biology but also provides strategic tools to overcome regeneration recalcitrance and enhance transformation efficiency in economically important crops.

References

Bao J, O'Donohue B, Sommerville KD, Mitter N, O'Brien C, Hayward A (2024) Tissue culture innovations for propagation and conservation of Myrteae-a globally important Myrtaceae tribe. Plants 13(16):2244

Barcaccia G, Palumbo F, Sgorbati S, Albertini E, Pupilli F (2020) A reappraisal of the evolutionary and developmental pathway of apomixis and its genetic control in angiosperms. Genes 11(8):859

Bettoni JC, Wang MR, Wang QC (2024) In vitro regeneration, micropropagation and germplasm conservation of horticultural plants. Horticulturae 10(1):45

Bhojwani SS, Dantu PK (2013) Cellular totipotency. In: Plant tissue culture: An introductory text. Springer, India, pp 63–74

Cao X, Yang H, Shang C, Ma S, Liu L, Cheng J (2019) The roles of auxin biosynthesis YUCCA gene family in plants. Int J Mol Sci 20(24):6343

Chong K (2025) Rewiring the stomatal lineage: a novel developmental pathway to plant totipotency. J Integr Plant Biol 67(12):3068

Dai H, Lv Y, Liu Z, Li L (2026) LEC2, a key player for cell totipotency in somatic embryogenesis and plant regeneration. Funct Integr Genomics 26(1):16

Hanning E (1904) Zur Physiologie pflanzlicher Embryonen. I. Ueber die Cultur von Cruciferen-Embryonen ausserhalb des Embrysacks. Bot Ztg 62:45–80

Ibanez S, Carneros E, Testillano PS, Perez-Perez JM (2020) Advances in plant regeneration: shake, rattle and roll. Plants 9(7):897

Jiang Y, Wei X, Zhu M, Zhang X, Jiang Q, Wang Z, Cao Y, An X, Wan X (2024) Developmental regulators in promoting genetic transformation efficiency in maize and other plants. Curr Plant Biol 40:100383

Kowalik S, Samoń M, Przyborowski M (2026) Molecular regulators of in vitro regeneration in wheat: roles of Morphogenic factors in transformation, genome editing, and breeding. Int J Mol Sci 27(3):1271

Long Y, Yang Y, Pan G, Shen Y (2022) New insights into tissue culture plant-regeneration mechanisms. Front Plant Sci 13:926752

Malabadi RB, Chalannavar RK, Kolkar KP (2025) Plant cell totipotency: plant tissue culture applications-An updated review. World J Adv Eng Technol Sci 16(02):112–135

Méndez-Hernández HA, Ledezma-Rodríguez M, Avilez-Montalvo RN, Juárez-Gómez YL, Skeete A, Avilez-Montalvo J, De-la-Peña C, Loyola-Vargas VM (2019) Signaling overview of plant somatic embryogenesis. Front Plant Sci 10:77

Morinaka H, Coleman D, Sugimoto K, Iwase A (2023) Molecular mechanisms of plant regeneration from differentiated cells: approaches from historical tissue culture systems. Plant Cell Physiol 64(3):297–304

Morończyk J, Brąszewska A, Wójcikowska B, Chwiałkowska K, Nowak K, Wójcik AM, Kwaśniewski M, Gaj MD (2022) Insights into the histone acetylation-mediated regulation of the transcription factor genes that control the embryogenic transition in the somatic cells of Arabidopsis. Cells 11(5):863

Pan C, Sretenovic S, Qi Y (2021) CRISPR/dCas-mediated transcriptional and epigenetic regulation in plants. Curr Opin Plant Biol 60:101980

Sirangi SK (2025) Tissue culture and conservation strategies for Indian Flora. Deep Science Publishing

Spinoso-Castillo JL, Bello-Bello JJ (2022) In vitro stress-mediated somatic embryogenesis in plants. In: Somatic embryogenesis: methods and protocols. Springer, New York, pp 223–235

Su YH, Tang LP, Zhao XY, Zhang XS (2021) Plant cell totipotency: insights into cellular reprogramming. J Integr Plant Biol 63(1):228–243

Wang FX, Shang GD, Wu LY, Xu ZG, Zhao XY, Wang JW (2020) Chromatin accessibility dynamics and a hierarchical transcriptional regulatory network structure for plant somatic embryogenesis. Dev Cell 54(6):742–757

Wang P, Si H, Li C, Xu Z, Guo H, Jin S, Cheng H (2025) Plant genetic transformation: achievements, current status and future prospects. Plant Biotechnol J 23(6):2034–2058

Wojcik AM (2025) LEC2 unlocks totipotency by unlocking chromatin. Trends Plant Sci 31:129–131

Wu M, Chen A, Li X, Li X, Hou X, Liu X (2024) Advancements in delivery strategies and non-tissue culture regeneration systems for plant genetic transformation. Adv Biotechnol 2(4):34

Xu P, Zhong Y, Xu A, Liu B, Zhang Y, Zhao A, Yang X, Ming M, Cao F, Fu F (2024) Application of developmental regulators for enhancing plant regeneration and genetic transformation. Plants 13(9):1272

Yuan HY, Kagale S, Ferrie AM (2024) Multifaceted roles of transcription factors during plant embryogenesis. Front Plant Sci 14:1322728

Zhu X, Xu Z, Wang G, Cong Y, Yu L, Jia R, Qin Y, Zhang G, Li B, Yuan D, Jin S (2023) Single-cell resolution analysis reveals the preparation for reprogramming the fate of stem cell niche in cotton lateral meristem. Genome Biol 24(1):194

Chapter 2
Developmental Regulatory Genes: The Molecular Switching

Abstract Plant cells exhibit outstanding developmental plasticity, enabling regeneration of whole plants through callus formation, organogenesis, and somatic embryogenesis. This regenerative capacity is governed by coordinated cellular and epigenetic reprogramming that re-establishes embryogenic competence in differentiated cells. These processes are regulated by developmental regulatory genes that function as molecular switches, including members of the *WUSCHEL* (*WUS*), *WUS-RELATED HOMEOBOX* (*WOX*), *BABY BOOM* (*BBM*), *LEAFY COTYLEDON* (*LEC*), *AGAMOUS-LIKE15* (*AGL15*), *SOMATIC EMBRYOGENESIS RECEPTOR KINASE* (*SERK*), and *SOMATIC EMBRYOGENESIS RECEPTOR-LIKE KINASE* (*KNOX*) families. These transcription factors interact with plant growth regulators, particularly auxin, and operate within complex gene regulatory networks that control meristem maintenance, cell fate transition, and embryo initiation. Along with developmental regulatory genes, epigenetic mechanisms such as DNA methylation dynamics, histone modifications, and chromatin remodeling further modulate transcriptional states during tissue culture-induced reprogramming. Manipulation of key morphogenic regulators has significantly improved in vitro regeneration and transformation efficiency, especially in recalcitrant genotypes. This chapter summarizes current knowledge on molecular switches that regulate the processes of totipotency, highlighting how developmental regulatory genes coordinate the shift from differentiated to embryogenic states in plants.

Keywords *AGL15* · *BBM* · Developmental regulatory genes · Growth hormones · *LEC* · Molecular regulators · *SERK* · Totipotency · *WUS*

2.1 Introduction

Plant cells possess remarkable plasticity, allowing them to regenerate entire organisms (i.e., callus somatic embryogenesis, shoot and root) from differentiated somatic cells, a phenomenon known as totipotency. Plant totipotency and somatic embryogenesis are generally based on two principles, i.e., cellular reprogramming and

P. Kumar, *Totipotency in Plants*, SpringerBriefs in Plant Science,
https://doi.org/10.1007/978-981-92-1631-4_2

epigenetic reprogramming. In cellular reprogramming plant cells retain the capacity for unlimited growth through the mitotic activity of meristematic cells, which continuously divide to renew themselves and produce cells for organ formation (Fambrini et al. 2022). This process is governed by a complex interplay of developmental regulatory genes, plant growth regulators, and epigenetic mechanisms. Key genes involved in initiating and maintaining totipotency, somatic embryogenesis, shoot induction, and callus formation include members of the WUSCHEL (*WUS*) and WUS-RELATED HOMEOBOX (*WOX*) transcription factor families, as well as other important regulators such as BABY BOOM (*BBM*), LEAFY COTYLEDON (*LEC*), and AGAMOUS-LIKE15 (*AGL15*). The induction of callus, somatic embryogenesis, root and shoot regeneration often involves a plant growth hormonal signal, such as auxin, that establishes an appropriate cellular environment. Stress and ectopic expression of embryo and meristem shows that transcription factors also play crucial roles (Su et al. 2021). Further, epigenetic reprogramming plays a vital role in re-establishing the competence of differentiated cells to undergo cellular dedifferentiation, acquire totipotency, and determine cell fate. Chromatin remodeling, involving post-transcriptional histone modifications and changes in DNA methylation, is necessary for cellular dedifferentiation and cell fate switching (Wójcikowska et al. 2020). Plant tissue culture is an extensively used technique in both industrial applications and basic and applied research for plant regeneration. In vitro regeneration methods include callus-mediated organogenesis, somatic embryogenesis, and shoot and root formation, leading to complete plantlets. The manipulation of culture conditions, such as the application of growth hormones, light regimes, and other environmental factors, has an important role in promoting organogenesis by regulating the expression of genes involved in regeneration processes (Long et al. 2022). The mechanisms that allow the preservation of the meristematic state of stem cells and ultimately help regulate cell fate are only partially understood. However, it has been shown that somatic embryogenesis, which is essential for the development of the embryo and the establishment and maintenance of the shoot apical meristem, can be accomplished by ectopic overexpression of specific genes (Horstman et al. 2017a, 2017b; Salaün et al. 2021). These genes include those encoding key regulatory transcription factors, such as the homeodomain protein *WUS*, AINTEGUMENTA-LIKE (*AIL*) family members, AP2/ERF-domain transcription factors PLETHORA4/BABY BOOM (*PLT4/BBM*) and PLETHORA5/EMBRYO MAKER (*PLT5/EMK*), the *MADS-box* transcription factor *AGL15*, the *NF-Y* family transcription factor *LEC1*, *B3-domain* transcription factors *LEC2* and FUSCA3 (*FUS3*), the Myeloblastosis (*MYB*) transcription factor *MYB118*, the Arginine-Tryptophan-Proline-Arginine-Lysine (*RWP-RK*) domain-containing protein RKD4/GROUNDED (*GRD*), the class I *KNOX* homeodomain protein SHOOT MERISTEMLESS1 (*STM1*), and the receptor-like kinase CLAVATA1 (*CLV1*). Additionally, other transcription factors exhibiting dynamic expression patterns during somatic embryogenesis have been identified through time-course analyses, including SOMATIC EMBRYOGENESIS RECEPTOR KINASE (*SERK*), CUP-SHAPED COTYLEDON1 (*CUC1*), and *CUC2*, as well as *STM*-like genes (Fambrini et al. 2022; Mehraj et al. 2025). The involvement of many different regulatory factors in the acquisition of embryogenic competence indicates that this process is controlled by several interconnected pathways. However, it is still unclear how these genes work together to initiate and maintain the formation of embryogenic cells (Farina et al. 2021). Growing evidence indicates that epigenetic modifications play a crucial role in determining cell fate by providing an additional regulatory layer that modulates the activity of numerous transcription factors involved in embryogenesis and organogenesis (Mehraj et al. 2025). Together with transcriptomic changes, these epigenetic mechanisms critically influence in vitro organogenesis and transformation efficiency. Moreover, the targeted manipulation of morphogenetic transcription factors, including the overexpression and ectopic expression of organogenesis-related genes, has been shown to enhance regeneration capacity and enable efficient transformation of recalcitrant plant genotypes. This chapter compiles current knowledge on developmental regulatory genes that function as molecular switches controlling embryogenic competence and cellular totipotency in plants. It mainly focuses on key classes of master regulators and transcription factors that initiate and maintain embryogenesis, their roles in chromatin remodeling, and the gene regulatory networks that stabilize and redirect developmental programs. By integrating molecular, genetic, and regulatory perspectives, this chapter provides a framework for understanding how developmental regulatory genes establish the transition between differentiated and totipotent states.

2.2 Transcriptional Changes of Developmental Regulatory Genes Through Plant Tissue Culture

Somatic embryogenesis involves a sequence of coordinated cellular and molecular reprogramming events as shown in Fig. 2.1. The molecular basis of somatic embryogenesis has been extensively investigated in model plants and major crops, including *Arabidopsis thaliana*, oil palm, maize, potato, soybean, and grapevine. More recently, substantial progress has also been made in understanding these mechanisms in woody and silvicultural species, such as spruce, pine, Japanese cedar, and other conifers (Mehraj et al. 2025). In plant tissue culture, wounding acts as a primary trigger for somatic cell dedifferentiation, largely mediated by the WOUND-INDUCED DEDIFFERENTIATION 1 (*WIND1*) gene (Méndez-Hernández et al. 2019). Several additional genes, including WOUND-INDUCED PROTEIN 1 (*WIP1*), ABA DEFICIENT 2 (*ABA2*), *AGL15*, *SERK1*, CHITINASE A1, and members of the GLUTATHIONE S-TRANSFERASE (*GST7*, *8*, *11*, *16*, and *19*) family, are also responsive to abiotic stress during somatic embryogenesis (Mehraj et al. 2025). Moreover, transcriptional regulation of key developmental regulators such as *LEC*, *WUS*, *FUS3*, ABA INSENSITIVE 3 (*ABI3*), and *AGL15* plays a crucial role in somatic embryogenesis across *Arabidopsis* and other plant species (Salaün et al. 2021). Shoot regeneration is governed by a conserved set of regulatory genes, including KNOTTED1 (*KN1*), *STM*, *WUS*, *CLV1*, *CLV2*, *CLV3*, *CUC1*, *CUC2*, and ENHANCER OF SHOOT REGENERATION (*ESR1*, *ESR2*) (Bidabadi and Jain

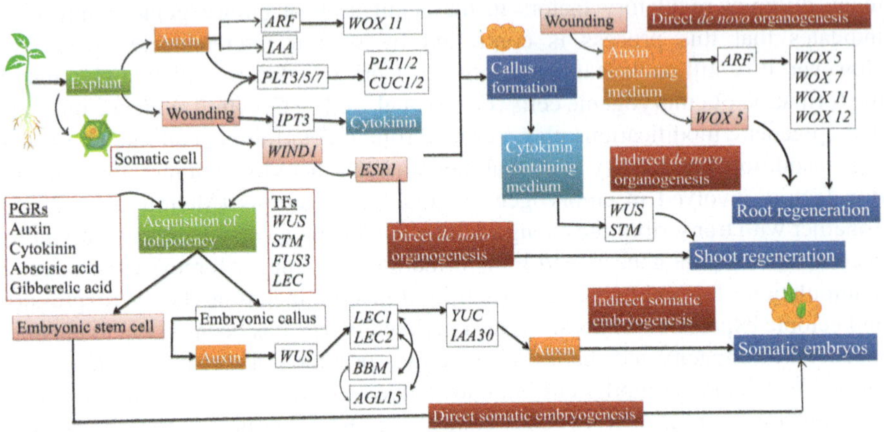

Fig. 2.1 Role of developmental regulatory genes in plant cell reprogramming and regeneration. (*AGL15 AGAMOUS-Like 15*, *ARF AUXIN RESPONSE FACTORs*, *BBM BABY BOOM*, *CUC1/2 CUP-SHAPED COTYLEDON1/2*, *ESR1 ENHANCER OF SHOOT REGENERATION*, *FUS3 FUSCA3*, *LEC LEAFY COTYLEDON*, *IAA* indole-3-acetic acid, *IPT3 isopentenyl transferase* 3, *L1L LEC1-LIKE*, *PLT PLETHORA*, *SERK SOMATIC EMBRYOGENESIS RECEPTOR-LIKE KINASE*, *STM SHOOT MERISTEMLESS*, *YUC YUCCA*, *WIND WOUND-INDUCED DEDIFFERENTIATION*, *WOX WUSCHEL-RELATED HOMEOBOX*, *WUS WUSCHEL*)

2020). During callus induction on callus induction medium, ETHYLENE RESPONSE FACTOR (*ERF*) transcription factors and *PLT* (*3*, *5*, *7*) are rapidly activated and subsequently induce *PLT1* and *PLT2*, leading to the formation of pluripotent, root-meristem-like callus tissue (Kerstens et al. 2024). In addition, *PLT3*, *PLT5*, and *PLT7* promote the expression of *CUC1* and *CUC2*, facilitating shoot meristem initiation (Lian et al. 2022). These findings demonstrate that plant tissue culture conditions profoundly alter the transcriptional landscape of multiple genes involved in in vitro organogenesis and regeneration (Table 2.1). Endogenous cellular responses triggered by in vitro developmental reprogramming activate intrinsic signaling pathways, resulting in species-specific transcriptional regulation of developmental regulatory genes. These developmental regulatory genes, in turn, activate, modulate, or constrain downstream targets involved in hormonal signaling, biosynthesis, and developmental pathways governing organogenesis in tissue culture. Conversely, successful cell proliferation and organ formation require precise genetic control and environmental coordination. Consequently, different plant growth hormones, controlled light conditions, and other environmental factors have been extensively examined to determine optimal culture conditions and factor combinations (Mehbub et al. 2022). Among these, growth hormones are most widely applied to externally modulate tissue culture responses and their influence on the transcriptional regulation of developmental regulatory genes.

Table 2.1 Developmental regulatory genes promoting in vitro regeneration in diverse plant species

Gene	Expression of gene	Species	Outcomes	References
AtWUS	Highly expressed	*Arabidopsis thaliana*	Formation of embryogenic calli and somatic embryos	Zuo et al. (2002)
	Overexpressed	*P. glauca*	Induction of somatic embryos and conversion into plantlets	Klimaszewska et al. (2010)
	Overexpressed	*Gossypium hirsutum*	Embryogenic callus formation	Bouchabke-Coussa et al. (2013)
AtWOX5, 11	Highly expressed	*Arabidopsis thaliana*	Initiation of callus formation	Hu et al. (2017)
AtWOX11, 12	Expressed	*Arabidopsis thaliana*	Promotes de novo root initiation	Liu JingChun et al. (2014)
MtWOX5	Expressed	*Medicago truncatula*	Helps in the formation of embryogenic lines and somatic embryos	Orłowska and Kępczyńska (2018)
MtWOX5, MtFUS3, MtWOX9, 11-like	Expressed	*Medicago truncatula*	Promotes formation of heart- and torpedo-stage somatic embryos	Kurdyukov et al. (2014)
WOX2, 9	Expressed	*Nicotiana tabacum*	Helps in early embryo formation	Zhou et al. (2018)
PpWOX2	Overexpression in somatic embryos	*P. pinaster*	Induction of callus and improve embryo quality	Hassani et al. (2022)
WOX1	Highly expressed	*Liriodendron* sp.	Plantlet formation	Long et al. (2023)
PaWOX2, 8/9	Expressed	*Picea abies*	Embryogenic stages	Mehraj et al. (2025)
VvWOX2, 3, 9, 11	Expressed	*Vitis vinifera*	Formation of somatic embryos	Mehraj et al. (2025)
AtLEC1, 2	Expression	*Arabidopsis thaliana*	Induction of somatic embryos and callus	Mehraj et al. (2025)
ZmLEC1	Maximum expression	*Zea mays*	Early stages of somatic embryos	Zhang et al. (2002)
AtLEC2	Ectopic expression	*Brassica napus*	Somatic embryo formation	Mehraj et al. (2025)
PkLEC1	High expression	*P. koraiensis*	Promotes somatic embryo formation from callus	Xu et al. (2025)
CcLEC1	Overexpression	*Coffea canephora*	Formation of globular and heart shape of somatic embryos	De-la-Peña (2013)

(continued)

Table 2.1 (continued)

Gene	Expression of gene	Species	Outcomes	References
TcLEC2	Expression	*Theobroma cacao*	Embryogenic callus formation	Shires et al. (2023)
MgSERK1	Expressed	*Melocactus glaucescens*	Shoot organogenesis	Torres-Silva et al. (2021)
CmSERK	Overexpression	*Cattleya maxima*	Embryogenic callus formation	Cueva-Agila et al. (2020)
HaSERK	Expressed	*Helianthus annuus*	Somatic embryo regeneration	Fambrini et al. (2006)
AtSERK1	Ectopic expression	*Arabidopsis thaliana*	Initiation of somatic embryos	Mehraj et al. (2025)
OsSERK1	Overexpression	*Oryza sativa*	Callus formation	Hu et al. (2005)
ZmSERK1, 2	Expression	*Zea mays*	Embryogenic callus formation	Mehraj et al. (2025)
AtBBM	Ectopic expression	*Arabidopsis thaliana*	Formation of somatic embryos	Mehraj et al. (2025)
ZmBBM	Co-expression	*Zea mays*	Somatic embryo formation	Mookkan et al. (2017)
CcBBM1	Expressed	*Coffea canephora*	Somatic embryo	De-la-Peña (2013)
BnBBM	Ectopic expression	*Brassica napus*	Somatic embryo formation	Mehraj et al. (2025)
CaBBM	Overexpression	*Capsicum annuum*	Embryo formation	Irikova et al. (2012)
AtWIND1, 2, 3, 4	Overexpression	*Arabidopsis thaliana*	Helps in callus and somatic embryo formation	Iwase et al. (2015)
ZmWIND1	Overexpression	*Zea mays*	Callus formation	Jiang et al. (2024)
AtAGL15	High expression	*Arabidopsis thaliana*	Somatic embryo formation	Pan et al. (2019)
GmAGL15	Expressed	*Glycine max*	Somatic embryos	El Ouakfaoui et al. (2010)
CsL1L	Expressed	*Citrus sinensis*	Somatic embryos and embryogenic callus formation	Tokuji et al. (2011)
AtL1L	Suppression	*Arabidopsis thaliana*	Defective embryos	Mehraj et al. (2025)

AGL15 AGAMOUS-Like 15, *BBM* BABY BOOM, *LEC* LEAFY COTYLEDON, *L1L* LEC1-LIKE, *SERK* SOMATIC EMBRYOGENESIS RECEPTOR-LIKE KINASE, *WIND* WOUND-INDUCED DEDIFFERENTIATION, *WOX* WUSCHEL-RELATED HOMEOBOX, *WUS* WUSCHEL

2.3 Classes of Developmental Regulatory Genes

2.3.1 WUS-Related Homeobox (WOX) Genes

Genes encoding homeobox proteins are characterized by a highly conserved 180 bp DNA sequence known as the homeobox, which translates into a 60-amino-acid globular homeodomain intricate for DNA binding (Bürglin and Affolter 2016). Evolutionary evidence suggests that homeobox genes originated early in eukaryotic evolution, originating from ancestral helix-turn-helix transcription factors (Fischer et al. 2022). In plants, the first homeobox gene was identified in the early 1990s through the characterization of the maize Knotted-1 (*Kn-1*) mutant, in which abnormal clusters of proliferating cells formed knot-like structures along lateral leaf veins due to sustained cell division (Vollbrecht et al. 1991). The *WOX* gene family encodes a group of plant-specific homeobox transcription factors that play important roles in regulating plant developmental processes, including embryogenesis, stem cell maintenance and organ formation. These functions are primarily associated with the regulation of cell proliferation and the prevention of premature differentiation (Li et al. 2020; Chen et al. 2023; Malabadi et al. 2025). In *Arabidopsis thaliana*, 15 *WOX* genes have been identified and classified into 3 major phylogenetic clades (*WUS*, INTERMEDIATE, and ANCIENT) based on sequence similarity and evolutionary relationships (Han et al. 2021). Functional specificity among *WOX* genes is largely determined by their ectopic expression patterns and interactions with other regulatory proteins. For example, *WUS* functions predominantly as a transcriptional repressor in shoot apical meristem stem cell populations while acting as a transcriptional activator during floral development by regulating *AGAMOUS* (*AG*) expression (Jha et al. 2020). *WOX* proteins contain a conserved helix–loop–helix–turn–helix (HLHTH) motif within the homeodomain, composed of three α-helices connected by a short loop and turn. The second α-helix mediates DNA binding through hydrogen bonding and hydrophobic interactions with exposed bases in the major groove of DNA. Structural analyses of sunflower (*Helianthus annuus*) *HaWUS* proteins have revealed a highly conserved tertiary structure comparable to homeodomains from other plant species (Riccucci et al. 2023). Members of the *WUS* clade possess a distinctive C-terminal *WUS*-box motif (T-LX-LFPXX), which is absent in other homeobox transcription factors and is essential for *WUS* function in stem cell regulation and floral development. The presence of conserved threonine and leucine residues at the beginning of the *WUS*-box distinguishes *WUS* clade members from *WOX* proteins in other clades, which exhibit amino acid substitutions at this position. Some *WOX* proteins also contain variable regions between the homeodomain and the *WUS*-box that may function as transcriptional activation domains, as well as a C-terminal *ERF*-associated amphiphilic repression (*EAR*) motif (Fambrini et al. 2022). The *EAR* motif, characterized by the consensus sequence (LVI)-X-(LVI)-X-(LVI), has been identified across members of all three *WOX* clades in *Arabidopsis* (Chow et al. 2023). In particular, the *EAR* domains of *WUS*, *WOX5* and *WOX7* are crucial for transcriptional repression in vegetative and floral meristems, partly through interactions with co-repressors such as TOPLESS (van der Graaff et al. 2009; Tvorogova et al. 2021). Notably, some rice (*Oryza sativa*) *WOX* proteins, including *WOX11* and *WOX3*, lack a canonical *EAR* motif yet function as transcriptional repressors. *WOX11* represses *OsRR2*, a negative regulator of cytokinin signaling, while *WOX3* suppresses *YABBY3* expression to control abaxial cell fate during leaf development (Zhao et al. 2015; Zhang et al. 2020a, 2020b, 2020c).

2.3.1.1 Roles of *WOX* Genes in Plants (In Vivo)

In *Arabidopsis thaliana*, *WUS* and *WOX5* are central regulators of stem cell maintenance in the shoot and root apical meristems, respectively (Sarkar et al. 2007). *WUS* expression is initiated at the 16-cell embryonic stage in inner apical cells and remains spatially restricted throughout embryogenesis, whereas *WOX5* is expressed during early root and cotyledon development. Loss of *WUS* or *WOX5* function leads to premature differentiation of totipotent stem cells in the shoot and root apical meristem, respectively, highlighting their conserved roles in shoot and root meristem. Beyond vegetative meristems, *WUS* also regulates ovule and anther development across multiple plant species (Wójcik et al. 2025). Several *WOX* family members coordinate embryogenesis and organ initiation. *WOX3* and its maize orthologs *NODULATION SIGNALING 1*/ *NODULATION SIGNALING 2* (*NS1*/*NS2*) promote founder cell recruitment during organogenesis, while *WOX6* prevents premature differentiation during integument formation and contributes to cold stress tolerance (Conklin et al. 2020; Dong et al. 2024). During embryonic patterning, *WOX2* is regulated by *WOX8* and *WOX9*, which is essential for shoot apical meristem formation, with multiple *WOX* genes acting redundantly during early development. As embryogenesis progresses, *WOX2* becomes restricted to the apical domain, whereas *WOX8* and *WOX9* localize to the basal region, collectively establishing embryonic polarity and localized auxin responses through interaction with PIN-FORMED1 (*PIN1*). *WOX9* further supports shoot apical meristem maintenance and cell proliferation during embryonic and post-embryonic development (Du et al. 2025; Ma et al. 2025). Members of the ANCIENT clade, *WOX13* and *WOX14*, are expressed in roots and floral organs, preventing premature differentiation (Du et al. 2025). Meristem homeostasis in the shoot apical meristem is maintained through a negative feedback loop between *WUS* and *CLV* signaling, whereby *WUS* induces *CLV3*, which restricts *WUS* expression via the *CLV1*/*CLV2* receptor complex (Agarwal et al. 2022; Wang and Jiao 2023). A comparable regulatory complex operates in roots, where *CLE40-ACR4* (CLAVATA3/ESR-Related 40/ *Arabidopsis* CRINKLY 4) signaling negatively regulates *WOX5* activity (Berckmans et al. 2020). *WUS* also modulates cytokinin signaling by repressing type-A *Arabidopsis* Response Regulator (*ARR-A*) genes. In floral meristems, *WUS* activates *AG*, which subsequently represses *WUS* to terminate stem cell activity (Pokimica et al. 2024). Recent studies have expanded *WOX* functions beyond meristem regulation. *LlWOX9* and *LlWOX11* promote bulbil formation in *Lilium lancifolium* through cytokinin-dependent regulation (He et al. 2021; Bai et al. 2025), while coordinated *WUS* expression accompanies shoot regeneration during asexual propagation in Crassulaceae species (Guo et al. 2015). *WOX13* plays a pivotal role in wound-induced callus formation and tissue repair by activating *WIND* genes (Ikeuchi et al. 2022). Additionally, *WUS* confers meristem resistance to cucumber mosaic virus by repressing ribosome-associated methyltransferases, thereby limiting viral protein synthesis (Wu et al. 2020). *WOX* genes also contribute to abiotic stress adaptation. Overexpression of *OsWOX13* enhances drought tolerance in rice, while differential regulation of *WOX* genes under drought, salinity, and cold stress has been reported in *Jatropha*, cotton, poplar, tea, and *Brassica napus*, highlighting their conserved roles in environmental responsiveness (Fambrini et al. 2022).

2.3.1.2 Activity of *WOX* Genes During Embryogenesis and Interaction with Plant Growth Hormones

Somatic Embryogenesis

Somatic cell reprogramming and the acquisition of embryogenic competence require elevated auxin levels, most commonly supplied as the synthetic auxin 2,4-dichlorophenoxyacetic acid (2,4-D). This compound is widely used to induce somatic embryogenesis and to initiate signaling cascades that modulate the expression of numerous somatic-embryogenesis-associated transcription factor genes (Wang et al. 2020). In several experimental systems, somatic embryogenesis induction is accompanied by a transient increase in endogenous indole-3-acetic acid (IAA) levels (Caeiro et al. 2022). Notably, the elimination of 2,4-D from the culture medium is often essential for embryo development from embryogenic callus. This removal activates the expression of *YUCCA2* (*YUC2*) and *YUC4*, which encode auxin biosynthetic enzymes, thereby increasing endogenous IAA accumulation (Suzuki et al. 2015; Garcia et al. 2019). Concurrently, somatic embryogenesis is associated with differential expression of genes encoding core components of the auxin signaling pathway, including AUXIN/INDOLE-3-ACETIC ACID (*Aux/IAA*) proteins and AUXIN RESPONSE FACTORs (*ARFs*). Live imaging analyses of *PIN* gene expression have demonstrated that uneven auxin distribution is critical for de novo morphogenesis during in vitro somatic embryogenesis (Wójcik et al. 2020). In *Arabidopsis thaliana*, an auxin gradient and *PIN1*-mediated polar auxin transport (PAT) are essential for the induction of *WUS* and the initiation of somatic embryogenesis. Moreover, several regulatory genes associated with zygotic embryogenesis that are upregulated during somatic embryogenesis induction, with *WUS* expression detected in embryogenic callus prior to the morphological appearance of somatic embryos (Fambrini et al. 2022). Further studies have shown that the expression patterns of key regulatory genes involved in root apical meristem formation correlates with the establishment of the embryonic root meristem during somatic embryogenesis in *Arabidopsis*. Early and partially overlapping expression of *WOX5* and *WUS* was observed in embryogenic callus before visible embryo formation. Additionally, cytokinin response signals were detected in specific regions corresponding to *WOX5* induction and subsequent somatic embryo formation. Overexpression of *ARR7* and *ARR15*, which act as negative feedback regulators of cytokinin signaling, disrupted root apical meristem initiation and inhibited somatic embryogenesis induction. These findings highlight the coordinated roles of auxin and cytokinin signaling in establishing apical basal polarity during somatic embryogenesis (Yamoune et al. 2021). In *Medicago truncatula*, comparative expression analyses of *WOX* and *PIN* genes during ovule development and somatic embryo progression demonstrated that *MtWOX11*-like and *MtPIN10* were highly expressed in ovules and activated during somatic embryogenesis, whereas other *WOX* and *PIN* genes showed low expression. These results suggest that specific regulatory network control early stages of both zygotic and somatic embryogenesis (Xu et al. 2024). In gymnosperms, such as *Picea abies*, *PaWOX2* expression peaks during the earliest stages of embryogenesis but is absent in non-embryogenic cultures, indicating its pivotal role in early somatic embryogenesis and its remarkable use as a molecular marker for embryogenic competence (Bueno et al. 2021). Furthermore, inhibition of PAT by N-1-naphthylphthalamic acid (NPA) disrupts embryo morphology and alters PIN1 expression during somatic embryogenesis, reinforcing the functional link between PAT and *WOX*-mediated embryo patterning in seed plants (Abas et al. 2021).

The interaction between plant growth hormones and *WOX* genes has been documented across multiple species. In *Triticum aestivum*, *TaWOX5* is predominantly expressed in roots and auxin- and cytokinin-induced calli, suggesting a role in root development and plant growth hormones-mediated regulation during somatic embryogenesis (Wójcik et al. 2025). In the interspecific hybrid *Liquidambar styraciflua* × *L. formosana*, transcriptomic analyses revealed activation of auxin biosynthesis and signaling pathways during somatic embryogenesis, with small auxin-upregulated RNA (*SAUR*) genes (Ai et al. 2023). Genes such as *PIN1*-like, *PIN2*, *AUX1*, *ARFs*, Gretchen Hagen 3 (*GH3*), and *SAUR* were markedly upregulated in embryogenic callus but downregulated during later somatic embryogenic stages, indicating their primary involvement in the transition from vegetative to embryogenic competence. Similarly, *WOX9* expression was specific to embryogenic callus and declined during somatic embryogenesis, whereas *WOX11* remained highly expressed throughout somatic embryo development and absent in non-embryogenic tissues, suggesting stage-specific regulatory functions (Qi et al. 2021). Time-course expression analyses have further elucidated the temporal regulation of *WOX* genes during somatic embryogenesis induction. In *Lactuca sativa*, *LsWUS1L* transcripts were scarcely detectable during the initial days of culture in the presence of auxin and cytokinin but increased progressively after 7–12 days, conferring with the initiation of adventitious structures and somatic embryos (Bull and Michelmore 2022). In contrast, *LsWUS2L* showed a rapid and significant increase in expression within the first four days of culture. This transcriptional activation coincided with the appearance of small, densely cytoplasmic cells with large nuclei, hallmarks of embryogenic competence, which subsequently organized into somatic embryos and adventitious buds. These observations suggest that *LsWUS2L* plays a critical role in the early induction phase of somatic embryogenesis (Bull 2022).

Zygotic Embryogenesis

Zygotic embryogenesis is regulated by a coordinated network of plant growth hormones, primarily auxin, abscisic acid (ABA), cytokinins, and gibberellins. Among these, auxin biosynthesis and the establishment of polar auxin transport are critical determinants of embryonic patterning and cell fate specification (Winnicki 2020). During early embryogenesis in *Arabidopsis thaliana*, auxin distribution is highly dynamic and spatially regulated. These processes depend on both auxin transport components and downstream response machinery (Hu et al. 2021). Directional auxin flow is mediated by members of the *PIN* family of auxin efflux carriers (Jha et al. 2025). Following the first zygotic division, *PIN7* localizes to the apical side of basal cell derivatives and subsequently reverses polarity at the globular stage. *PIN1* exhibits polarized localization from the mid-globular stage onward, corresponding to an early auxin maximum in the apical domain that later shifts to the basal pole during the globular stage. At later stages, *PIN4* localizes to the hypophysis and adjacent suspensor cells, further refining auxin distribution patterns (Adamowski and Friml 2015). Disruption of *PIN* polarity, as observed in *GNOM* mutant embryos, results in defective embryonic patterning due to impaired *PIN* recycling between the plasma membrane and endosomal compartments. Severe disruption of *PIN*-mediated transport alters auxin responses in apical regions, leading to aberrant cell fate transitions, including cotyledon-to-root development (Dhonukshe et al. 2008). *WOX* transcription factors function in parallel with auxin signaling to establish embryonic axes and lineage specification. In *Arabidopsis*, *WOX2* and *WOX8*, initially co-expressed in the zygote, subsequently exhibit lineage-specific expression in the apical and basal domains, respectively, thereby defining the apical basal axis. *WOX2* expression in the apical lineage is an early downstream outcome of *WOX8*/*WOX9* activity and is closely associated with localized auxin responses in the proembryo. While *WOX8* expression is largely auxin-independent, both *WOX8* and *WOX9* are activated in the zygote, basal daughter cell and hypophysis by the zinc-finger transcription factor *WRKY2* (Huang and Sun 2025). Phylogenetic and functional analyses suggest that this *WOX-PIN*–auxin regulatory complex represents an evolutionarily conserved mechanism governing embryo patterning in seed plants (Verma et al. 2021). Comparative studies in monocots indicate species-specific variations in *WOX*-mediated embryogenesis. In maize, expression profiling suggests that apical–basal lineage specification may involve distinct *WOX* regulators or occur at later developmental stages compared with *Arabidopsis* (Luo et al. 2024). In gymnosperms, such as *Picea abies*, *PaWOX2* plays a pivotal role during early embryogenesis, particularly in protoderm formation and suspensor development. Early downregulation of *PaWOX2* significantly reduces the production of mature embryos, whereas its suppression at later stages does not affect embryo maturation, underscoring its stage-specific function during early zygotic development (Zhu et al. 2016).

2.3.1.3 Activation and Shoot Regeneration of WOX Genes

In *Arabidopsis thaliana*, de novo shoot apical meristems can be regenerated from cultured root explants. Using fluorescent reporter lines, they characterized the early patterning events underlying shoot meristem formation and identified a small population of progenitor cells that initiate shoot apical meristem development through a sequence of gene expression events. This process involves the coordinated activity of key regulators, including *CUC2*, *WUS*, *PIN1*, *STM*, *FILAMENTOUS FLOWER* (*FIL*), *REVOLUTA* (*REV*), *ARABIDOPSIS THALIANA MERISTEM LAYER 1* (*ATML1*), and *CLV3* (Xue et al. 2020; Yang et al. 2023). These genes, along with auxin and cytokinin signaling, regulate successive stages of adventitious meristem initiation and coordinated development. Notably, *WUS* activity was shown to be essential for de novo shoot apical meristem formation, highlighting its functional interaction with cytokinin signaling pathways (Pokimica et al. 2024). This regulatory framework was further demonstrated by Dai et al. (2017), through chromatin immunoprecipitation and transient expression assays that *ARR12* directly binds to the *WUS* promoter in in vitro cultured plant cells. These findings show *ARR12* as a molecular integrator linking cytokinin signaling to *WUS* expression during shoot regeneration. Among in vitro regeneration strategies, the induction of adventitious buds via an indirect pathway involving an intermediate callus phase is one of the most widely employed approaches. The initiation of this process is strongly dependent on the crosstalk between auxin and cytokinin signaling pathways. Zhai and Xu (2021) reported that, in *Arabidopsis*, the tissue architecture formed following callus induction closely resembles that of root primordia or apical meristems. Within this structure, a middle cell layer exhibiting a quiescent center–like transcriptional identity was identified as the site of pluripotency acquisition and organ regeneration competence. At this site, *WOX5* directly interacts with *PLT1* and *PLT2* to activate the expression of *TRYPTOPHAN AMINOTRANSFERASE OF ARABIDOPSIS1* (*TAA1*), thereby promoting endogenous auxin biosynthesis. Concurrently, *WOX5* interacts with the B-type cytokinin response regulator *ARR12* and represses A-type *ARR* genes, effectively relieving negative feedback within the cytokinin signaling pathway. Together, enhanced auxin production and increased cytokinin sensitivity enable the attainment of pluripotency in the middle cell layer of the callus, ultimately facilitating organ regeneration (Šmeringai et al. 2023).

2.3.2 LEC *and LEC1-LIKE (*L1L*) Genes*

The involvement of *LEC* and *FUS3* genes in embryogenesis has been demonstrated across numerous plant species, where their ectopic expression is adequate to induce somatic embryogenesis in the absence of exogenous growth hormones in culture media (Tang et al. 2017). The expression of *LEC* and *FUS3* plays a central role in regulating hormone biosynthesis and signaling pathways that direct in vitro organogenesis. *LEC*-dependent in vitro embryogenesis has been reported in a wide range of species, including *Arabidopsis thaliana*, *Theobroma cacao*, *Brassica napus*, *Nicotiana tabacum*, *Ricinus communis*, *Zea mays*, *Capsicum annuum*, *Medicago truncatula*, *Larix decidua*, *Manihot esculenta*, *Phalaenopsis* spp., *Daucus carota*, and *Gossypium hirsutum* (Mehraj et al. 2025). In particular, *AtLEC1*, *AtLEC2*, and *AtFUS3* have been identified as key regulators of in vitro embryogenesis and somatic embryogenesis (Barreto et al. 2019). The expression pattern of *ZmLEC1* during somatic embryogenesis closely resembles with *AtLEC1* during zygotic embryo development, suggesting that somatic embryogenesis in maize shows similarity with zygotic embryogenic program (Zhu et al. 2018). In *Arabidopsis*, *AtLEC1* regulates both zygotic and somatic embryogenesis through its involvement in transcriptional networks governing hormone signaling particularly auxin pathways as well as light signaling (Wójcik et al. 2020). During seed germination, somatic embryogenic competence is lost, partly due to auxin-induced degradation of the repressive histone mark *H3K27me3* in AP2 and B3 transcription factor loci. Conversely, auxin signaling becomes reactivated during embryonic maturation termination. Similar auxin-mediated epigenetic regulation has been observed in *Triticum aestivum*, where degradation of *H3K27me3* at *TaBOF5.6 and TaBOF3.4* leads to gene upregulation and enhanced callus formation (Liu et al. 2023). Functional activation of AP2 and B3 family transcription factors, including *BBM*, *LEC*, *ABI3* and *FUS3*, is required for the initiation of somatic embryogenesis. Notably, their transcription is generally absent in three-day-old seedlings cultured in vitro, with *FUS3* remaining undetectable at this stage (Yuan et al. 2024). Among these factors, *LEC2* exhibits rapid transcript accumulation in seedlings and directly activates *WOX2* and *WOX3*, thus promoting somatic embryogenesis induction. Analysis of single, double, and triple *wox* mutants further demonstrated that although *WOX2* and *WOX3* are necessary for somatic embryogenesis, they are not sufficient on their own (Wang et al. 2020). The capacity of *LEC1* to induce embryogenesis is species-dependent. While *LEC1* alone fails to induce callus, somatic embryos, or shoot regeneration in hormone-free media in *N. tabacum*, it independently promotes embryo formation, embryogenic callus, or embryogenic cell development in species such as *C. annuum*, *G. hirsutum*, *Coffea canephora*, *D. carota*, *Z. mays*, *Pinus koraiensis*, *M. esculenta*, and *Phalaenopsis aphrodite*. Moreover, *LEC1* expression increases progressively with the duration of embryogenic induction (Mehraj et al. 2025).

Genetic analyses have revealed that *lec1*, *lec2*, and *fus3* mutants exhibit abnormal cell division patterns and defective suspensor development during globular, heart, and torpedo stages of somatic embryogenesis, whereas *abi3* mutants do not display comparable abnormalities. In *Arabidopsis*, these mutations dramatically reduce somatic embryo formation rates relative to wild-type plants, underscoring the essential role of *LEC* and *FUS3* in somatic embryogenesis induction. Mutant analyses further indicate that *LEC* genes function early in embryogenesis, as certain double and triple mutant combinations completely stop somatic embryo formation (Horstman et al. 2017a, 2017b). Overexpression of *LEC2* enhances somatic embryogenesis, promotes improved germination, and rescues defective cotyledon phenotypes in *lec2* mutants, resulting in callus-like growth and somatic embryo formation.

The application of exogenous 6-benzyladenine (6-BA) and α-naphthaleneacetic acid (NAA) further increases shoot regeneration in *LEC2*-overexpressing *Arabidopsis*, an effect that is amplified by co-overexpression of ISOPENTENYL TRANSFERASE (*IPT*) genes (Mehraj et al. 2025). The *L1L* gene functions cooperatively with LEC1 during embryogenesis. Suppression of *L1L* in *Arabidopsis* results in embryonic defects distinct from those observed in *lec1* mutants, indicating a partially independent regulatory role (Orłowska et al. 2017). *L1L* expression has also been detected during embryogenesis and somatic embryogenesis in *Helianthus annuus*, *T. cacao*, and *Camellia sinensis*, further supporting its conserved involvement in embryogenic processes (Garcia et al. 2018). In *Medicago truncatula*, cultivar-specific expression of *LEC1* and *L1L* correlates with embryogenic competence. The embryogenic line (M9–10a) produces embryos, whereas the non-embryogenic line (M9) forms callus without embryo development after 10 days of in vitro culture. At this stage, *MtLEC2*, *MtFUS3*, and *MtABI3* transcripts accumulate and increase rapidly by day 14, indicating conserved transcriptional regulation during somatic embryogenesis induction. *LEC2* promotes embryogenic cell and somatic embryo formation by activating auxin-responsive genes (*IAA1*, *IAA17*, *ACS4*), auxin biosynthesis genes (*YUC1*, *YUC2*, *YUC4*, *YUC10*), auxin transport genes (*PIN7*), and auxin signaling components (SHY2), as well as *FUS3* and *ABI3* (Barreto et al. 2019). *FUS3* and *ABI3* regulate embryogenesis by targeting *AGL15*, thereby modulating abscisic acid, gibberellic acid, and ethylene biosynthesis pathways (Tian et al. 2020; Joshi et al. 2022). Additionally, *FUS3* interacts with LEC proteins to transcriptionally regulate *YUC4*, a gene involved in root development (Tang et al. 2017). Collectively, *LEC* and *L1L* genes exert broad control over hormonal pathways, particularly auxin and ABA metabolism, and form part of the LAFL regulatory network (*LEC1*, *ABI3*, *FUS3*, and *L1L*), which represents a central hub in embryogenesis. Consequently, ectopic expression of LAFL components in vegetative tissues induces cellular reprogramming toward embryonic developmental pathways and significantly enhances in vitro plant regeneration efficiency.

2.3.3 **SERK** *Genes*

The involvement of *SERK* in somatic embryogenesis was first reported in *Daucus carota* and subsequently confirmed in *Arabidopsis thaliana* (Shah et al. 2001). *SERK* expression is often tissue-specific and regulated by plant growth hormones present in culture media. Different experiments demonstrate that *SERK* plays an active role in callus induction, somatic embryogenesis, and shoot and root organogenesis across a wide range of plant species. *SERK* has a preferential role in the regulation of somatic embryogenesis, i.e., stage- and cell-type-specific expression patterns have been documented in *Passiflora edulis* and *Araucaria angustifolia*, respectively (Phong et al. 2023). In embryogenic cell suspension cultures of *Musa acuminata*, variation in somatic embryo formation frequency was positively correlated with the expression level of *MaSERK1*, indicating that *SERK* expression is closely associated with embryogenic competence. Moreover, *MaSERK1* transcript abundance was influenced by subculture duration, further emphasizing the dynamic regulation of *SERK* during in vitro culture (Adero et al. 2023). In orchids, *SERK1* expression was significantly higher in reproductive tissues than in vegetative organs, reinforcing its proposed function in embryogenesis-related processes. Functional studies involving the expression, suppression, and overexpression of *SERK* genes across diverse plant systems consistently demonstrate their essential involvement in in vitro organogenic and embryogenic pathways (Kowalik et al. 2026). The role of *SERK* in embryogenesis-associated signal transduction cascades has also been characterized in *Poa pratensis*, where *SERK* was implicated in early cellular signaling events underlying embryogenic induction. Furthermore, in *Coffea canephora*, *CcSERK1* was shown to activate auxin biosynthesis (via *CcYUC1* and *CcYUC2*) and polar auxin transport (via *CcPIN1* and *CcPIN4*), leading to the subsequent activation of *CcWUS*, *CcAGL15*, and *CcBBM*, thereby promoting somatic embryogenesis (Mehraj et al. 2025). These conclusions indicate that *SERK* genes are preferentially expressed in cells competent for somatic embryogenesis and that elevated *SERK* expression is a critical prerequisite for the successful induction and progression of somatic embryogenesis.

2.3.4 **BBM** *Genes*

BBM is a key regulator of in vitro regeneration efficiency and somatic embryogenesis across various plant species. *BBM* expression has been detected in embryogenic cells, callus, protocorm-like bodies, and somatic embryos, where it actively regulates in vitro organogenesis (Li et al. 2025). Constitutive expression of *Glycine max BBM1* (*GmBBM1*) in transgenic *Arabidopsis thaliana* induces somatic embryogenesis, mediated by a *BBM*-specific cis-regulatory sequence motif (*bbm-1*) that functions synergistically with the eudicot ANT (euANT2) motif (Zhou et al. 2023; Yavuz et al. 2024). Orthologs closely related to *GmBBM* and *MtBBM* in *Rosa canina* (*RcBBM1* and *RcBBM2*) also include the *bbm-1* motif and have been involved in callus formation and protocorm-like body development (Yang et al. 2014). In wheat *bbm-1* motif copies have been mapped to chromosomes 6AL, 6BL, and 6DL; nevertheless, only the A and D genome copies are functional in embryogenic microspore populations. This finding suggests that *TaBBM* contributes to enhanced doubled haploid production, tissue culture responsiveness, and transformation efficiency (Bilichak et al. 2018). *BBM* acts upstream of the *LAFL* regulatory network to stimulate auxin biosynthesis via the *YUCCA* gene family, thereby promoting somatic embryos initiation in *Arabidopsis* and *Brassica napus* (Li et al. 2022). By elevating endogenous auxin levels and activating auxin biosynthetic pathways, *BBM* induces cellular reprogramming that is essential for somatic embryogenesis. Moreover, *BBM* plays a pivotal role in establishing the appropriate balance between endogenous auxin and cytokinin required for cell proliferation during plant regeneration (Li et al. 2025). *BBM* expression exhibits developmental stage specificity during embryogenesis. In *Capsicum annuum* (chili/red pepper), *LEC1* and *BBM* are expressed at high levels throughout somatic embryo development, whereas in *Theobroma cacao*, *BBM* expression is temporally regulated during somatic embryogenesis (Irikova et al. 2012; Florez et al. 2015). Although *BBM* overexpression induces somatic embryos in *Populus tomentosa* and *T. cacao*, subsequent embryo development is impaired in cocoa, indicating that *BBM* primarily functions in somatic embryos initiation rather than in later developmental stages (Xu et al. 2024). Ectopic or constitutive expression of *BBM* significantly enhances in vitro organogenesis in multiple species, including *A. thaliana* (*TcBBM* and *RcBBM*), *B. napus* (*BnBBM*), *T. cacao* (*TcBBM*), and *P. tomentosa*. Notably, somatic embryos induced by *TcBBM* overexpression in *A. thaliana* occur independently of exogenous hormone application, whereas in *T. cacao*, hormone supplementation remains essential (Garcia et al. 2018; Xu et al. 2024). Furthermore, co-expression of *ZmBBM* (*Zea mays*) and *ZmWUS* synergistically improves somatic embryogenesis efficiency in maize and sorghum (*Sorghum bicolor*) (Duan et al. 2022). These findings demonstrate that *BBM* functions through direct or indirect activation of auxin biosynthesis genes and modulation of auxin-cytokinin homeostasis, thereby facilitating somatic embryogenesis and enabling regeneration programs as shown in Fig. 2.1.

2.3.5 AGL15, ABI, AHL, *and* PLT *Genes*

Overexpression of *AGL15* in *Arabidopsis thaliana* and *Glycine max* significantly enhanced somatic embryogenesis efficiency, although this effect diminished over prolonged culture periods (Joshi et al. 2022). *AGL15* primarily regulates genes involved in gibberellin and ethylene metabolism and additionally modulates auxin signaling through the activation of gibberellin 2-oxidase 6 (*GA2ox6*) and *IAA30* (Zheng et al. 2016). *ABSCISIC ACID INSENSITIVE* (*ABI*) genes play a crucial role in in vitro root regeneration from callus tissue. Specifically, *ABI3* and *ABI4* are key regulators of in vitro embryogenesis. In *abi3* mutants, reduced expression of *AUX1* and *GH3* resulted in a complete failure of root regeneration, indicating that the absence of *ABI3* disrupts auxin homeostasis and its spatial distribution during in vitro root formation. In contrast, *ABI3* overexpression did not promote somatic embryogenesis, likely due to insufficient activation of auxin-responsive genes (Sengupta and Nag Chaudhuri 2020).

AT-HOOK MOTIF CONTAINING NUCLEAR LOCALIZED (*AHL*) transcription factors are also implicated in somatic embryogenic regulation. *AHL15* is expressed during hormone-independent somatic embryogenesis in *Arabidopsis*, while the upregulation of *AHL15*, *AHL19*, *AHL20*, and *AHL29* in the presence of 2,4-D further stimulated somatic embryo induction (Karami et al. 2021; Tang et al. 2023). Although the *ahl15* single mutant did not exhibit a marked reduction in somatic embryos under 2,4-D treatment, the *ahl15 ahl19 amiRAHL20* triple mutant showed a significant decrease in somatic embryo formation, highlighting the functional redundancy and collective importance of multiple *AHL* genes in 2,4-D-mediated somatic embryogenesis. Especially, overexpression of *AHL* genes acts downstream of *BBM* and enhances *BBM*-induced somatic embryogenesis efficiency (Karami et al. 2021).

The *PLETHORA* (*PLT*) gene family, particularly *PLT3*, *PLT5*, and *PLT7*, regulates two key developmental processes and is essential for de novo shoot regeneration in *Arabidopsis* through the modulation of *CUC2* expression (Rodríguez-Herrera et al. 2025). Loss-of-function mutations in these genes result in callus formation and intermediate cell masses that lack competence for shoot regeneration (Cao et al. 2024). In conclusion, *AGL15*, *ABI*, *AHL*, and *PLT* genes play integral roles in in vitro organogenesis by modulating hormone biosynthesis, signaling pathways, or the expression of hormone-related genes, thereby coordinating cellular reprogramming and regenerative competence.

2.3.6 **WIND1** *Gene*

Plant cells are capable of responding to wounding in culture media lacking exogenous plant growth hormones by initiating callus formation at wound sites, a process primarily mediated by endogenous auxin signaling. Although wound-induced callus and plant growth hormonal-induced callus share certain developmental features, they differ histologically. Specifically, wound-induced callus lacks the expression of root meristem markers, whereas callus induced by growth hormones exhibits root meristem identity; nevertheless, both callus types undergo cellular dedifferentiation that enables further developmental progression (Ikeuchi et al. 2017). The transcriptional activation of *WIND* genes (*WIND1*, *WIND2*, *WIND3*, and *WIND4*) plays a vital role in regulating callus induction following wounding, a process accompanied by enhanced cytokinin biosynthesis and signaling (Iwase et al. 2021). Overexpression of *AtWIND1–4* promotes callus formation in hormone-free culture media in *Arabidopsis thaliana*, *Brassica napus*, *Nicotiana tabacum*, and *Solanum lycopersicum*. Comparative analyses of wild-type and *WIND1*-overexpressing Arabidopsis lines under wounded and unwounded conditions demonstrated that shoot regeneration from roots did not occur in the absence of wounding, even in cytokinin-rich shoot induction media, underscoring the essential role of wound-derived signals in de novo shoot organogenesis (Ribeiro et al. 2024). Consistently, transgenic *B. napus* (*35S::AtWIND1-GR*) and tomato (*35S::AtWIND1-GR*) lines were capable of producing callus and subsequent shoots in hormone-free culture media. *WIND* transcription factors modulate endogenous hormone levels and reprogram metabolic pathways and gene expression to facilitate callus formation at wound sites (Xu et al. 2024). In particular, *AtWIND1* directly upregulates *ESR1*, thereby promoting callus formation (Fig. 2.1) and shoot regeneration in *Arabidopsis* and *B. napus*. Functional analyses revealed that esr1 mutants exhibited defective callus and shoot development, whereas *ESR1* overexpression enhanced both processes. Moreover, significant downregulation of *ESR1* was observed in *ProWIND1::WIND1-SRDX* lines, and ectopic *ESR1* overexpression in this background failed to rescue normal callus and shoot development, further confirming that *ESR1* operates downstream of *WIND1* (Kshetry et al. 2025). These findings indicate that while the primary function of *WIND1* is the induction of callus formation in response to wounding, it also contributes to subsequent shoot and root regeneration through transcriptional activation of *ESR1*. Thus, the *WIND1-ESR1* regulatory complex constitutes a critical pathway underlying wound-induced in vitro organogenesis.

2.4 Developmental Regulatory Genes Contribute to Plant Transformation Efficiency

Plant transformation efficiency refers to the successful integration of foreign DNA into plant cells, followed by the regeneration of transformed cells into organs or complete plants through in vitro organogenesis. Limited regenerative capacity remains one of the principal constraints in achieving efficient plant transformation. Transformation efficiency is largely species-dependent; nonetheless, several studies have demonstrated that both transformation and regeneration efficiencies can be significantly enhanced, thereby reducing species-specific limitations (Wang et al. 2025). The manipulation of key developmental regulatory genes involved in oallus formation, somatic embryogenesis, and protocorm-like body development has proven effective in improving plant transformation efficiency (Xu et al. 2024). In particular, *Agrobacterium*-mediated transformation of developmental regulatory genes markedly enhances both transformation success and subsequent regeneration capacity across diverse plant species. For example, the introduction of the *AtWUS* gene into multiple plant species significantly promotes transformation efficiency and improves the regeneration of transformed tissues during in vitro culture (Lou et al. 2022; Zhang et al. 2023).

The small signaling peptide *REGENERATION FACTOR 1* (*REF1*) has also been shown to enhance transformation and regeneration efficiency in several species by activating *WIND1* through *PEPR1/2 ORTHOLOG RECEPTOR-LIKE KINASE 1* (*REF1–PORK1*)-mediated signaling (Schaller 2024; Wang et al. 2025). As a result, transformation and regeneration efficiencies increased substantially in *Triticum aestivum* (eight-fold and four-fold), *Solanum lycopersicum* (6–12-fold and 5–19-fold, respectively), *Zea mays* (four-fold and six-fold), and *Glycine max* (five-fold and nine-fold) (Mehraj et al. 2025). The inducible expression of *AtWUS* and *AtLEC2* in *Nicotiana tabacum* (*XVE::WUS* and *XVE::LEC2* constructs) demonstrated distinct regeneration outcomes. *XVE::WUS* lines regenerated morphologically normal plantlets in both the presence and absence of β-estradiol, whereas *XVE::LEC2* lines produced plantlets exhibiting abnormal phenotypes (Rashid et al. 2007). In *Picea mariana*, β-estradiol treatment increased the transcript levels of *CHAP3A* (a conifer homolog of *LEC1*) by two- to three-fold depending on concentration and culture duration. However, *CHAP3A* expression did not alter growth patterns, phenotypes, and somatic embryogenesis formation in either transgenic or non-transgenic plants. Similarly, β-estradiol treatment elevated *AtWUS* transcript levels up to six-fold within six days, followed by a decline to baseline levels by day 13. Although early embryo morphology remained unaffected, somatic embryogenesis frequency varied depending on β-estradiol presence and transgene expression (Klimaszewska et al. 2010). Subsequent embryo maturation on ABA-supplemented media revealed that excessive β-estradiol (2.0 μM) led to approximately 70% abnormal seedlings, whereas lower concentrations (0.02 μM) or its absence resulted in predominantly normal seedling development. Overexpression of *TaWOX5* in *T. aestivum* significantly increased transformation efficiency while reducing dependency on genotype and transformation method (Chang et al. 2024). In tobacco, *XVE::WOX2* transgenic lines exhibited enhanced callus proliferation from leaf explants in β-estradiol-supplemented liquid MS medium compared to *XVE::WOX8* and *XVE::WOX9* lines. Hybrid lines (*XVE::WOX2* × *XVE::WOX8* and *XVE::WOX2* × *XVE::WOX9*) displayed vigorous growth and root regeneration on solid MS medium supplemented with β-estradiol, whereas the parental lines failed to regenerate (Kyo et al. 2018). Co-transformation strategies involving *BBM* and *WUS* genes have further demonstrated substantial improvements in transformation efficiency across both monocotyledonous (rice, wheat, maize, sorghum, sugarcane) and dicotyledonous species (tomato, pepper, bok choy, snapdragon). A maize-optimized *ZmBBM/WUS2* construct enabled embryogenic callus formation and successful plant regeneration in a wide range of cereals and grasses, including rice, barley, maize, sorghum, foxtail millet, rye, pearl millet, switchgrass, and teff (Mehraj et al. 2025). Additionally, the transformation of *GROWTH-REGULATING FACTOR* (*GRF*) genes has been shown to enhance transformation efficiency in *Brassica napus*, *Lactuca* spp., *Glycine max*, *Zea mays*, *Citrullus lanatus*, *Helianthus annuus*, *Cannabis sativa*, and *Beta vulgaris* ssp. *vulgaris*. In particular, the *TaGRF4–GIF1* complex significantly improved transformation efficiency in *Triticum turgidum* var. *durum*, *T. aestivum*, *Oryza sativa*, and *T. rimpaui* (Xu et al. 2024). Similarly, overexpression of *PLT5* enhanced transgenic plant regeneration in *Solanum lycopersicum*, *Brassica rapa* and *Capsicum annuum* (Lian et al. 2022). Two additional transcription factors, *TaDOF5.6* and *TaDOF3.4*, have recently been identified as positive regulators of transformation efficiency in wheat. In *Arabidopsis thaliana*, *cuc1–1*, *cuc2–1*, and *cuc1–1 cuc2–1* mutants exhibited reductions of 50%, 50%, and 90%, respectively, in adventitious shoot formation from callus compared with a 15% reduction in wild-type plants after 10 days on shoot induction medium. These mutants displayed normal sensitivity to auxin and cytokinin, whereas overexpression of *CUC* genes promoted adventitious shoot formation (Liu et al. 2023). Transformation with *ESR2*-induced cytokinin-independent callus formation and *ESR2* overexpression altered the expression of A-type *ARR* genes (*ARR4*, *ARR5*, *ARR6*), suggesting a role in regulating cell division. Conversely, *ESR2* knockout reduced *CUC1* expression, resulting in decreased shoot regeneration and abnormal cotyledon development (Long et al. 2022).

Finally, overexpression of *GmAGL15* in soybean increased somatic embryo formation twofold in 2,4-D-supplemented media (Paul et al. 2022). Introduction of the nitrite reductase (*NiR*) gene from the Kasalath rice cultivar into *Oryza sativa* cv. Koshihikari calli promoted prolific shoot regeneration, and *NiR*-overexpressing constructs enabled normal growth even under high-nitrite conditions in Nipponbare cells (Zhang et al. 2020a, 2020b, 2020c). Collectively, these findings conclude that the targeted manipulation of growth and development associated genes is an effective strategy for enhancing plant transformation efficiency and in vitro organogenesis across a wide range of species.

2.5 Molecular Regulators of Plant Totipotency

The regenerative potential of plants arises from the inherent properties of cellular totipotency and pluripotency, which play a fundamental role in plant tissue regeneration. Under conditions such as stress, wounding, or hormonal stimulation, isolated plant cells cultured in vitro can initiate the formation of new tissues or organs. This phenomenon demonstrates the induction of cellular pluripotency under ex vivo conditions (Ikeuchi et al. 2016). Understanding the molecular mechanisms governing plant regeneration is therefore essential for advancing both fundamental plant biology and its agricultural applications. The molecular processes underlying de novo organogenesis (including root and shoot regeneration) and somatic embryogenesis are illustrated in Fig. 2.1. At the site of wounding in plants, or on explants cultured in vitro, a mass of rapidly proliferating cells known as callus is formed (Xu et al. 2024). During de novo organogenesis, the callus formed is typically non-embryogenic, exhibiting pluripotency and the capacity to regenerate roots and shoots. In contrast, embryogenic callus is associated with somatic embryogenesis and reflects the totipotent nature of plant cells (Ikeuchi et al. 2013). The formation of non-embryogenic callus can be stimulated through the application of auxin and cytokinin, or by environmental factors such as wounding and other severe stress conditions. Callus tissues developed on callus-inducing medium (CIM) exhibit histological characteristics similar to those of root meristematic tissues. During this process, auxin promotes the degradation of IAA14, leading to the activation of the transcription factors *ARF7* and *ARF19*. These transcription factors directly induce the expression of LATERAL ORGAN BOUNDARIES DOMAIN (*LBD16*, *LBD17*, *LBD18*, and *LBD29*), which collectively promote callus formation (Xu et al. 2018). *LBD* proteins subsequently activate the expression of numerous genes involved in cell proliferation and cell wall modification. In *Arabidopsis*, *LBD* transcription factors interact with the BASIC REGION/LEUCINE ZIPPER MOTIF 59 (*bZIP59*) transcription factor to form a transcriptional complex that co-regulates growth factor-induced callus formation. Furthermore, the *LBD16–bZIP59* complex directly regulates the expression of the FAD-binding Berberine (*FAD-BD*) gene involved in cell wall component metabolism. In addition, *LBD18* regulates the expression of E2 PROMOTER BINDING FACTOR a (*E2Fa*), a key transcription factor controlling the cell cycle, and also activates genes responsible for cell wall relaxation, indicating that dynamic remodeling of the cell wall plays an important role in callus development. Wounding further activates cytokinin signaling by stimulating cytokinin biosynthesis and through *WIND1*-mediated pathways. *ARR*-mediated cytokinin signaling promotes the re-entry of cells into the cell cycle through the activation of *CYCD3* (*Cyclin D*) (Xu et al. 2024). Moreover, wounding induces the expression of *PLT* genes, which are essential for callus formation. When explants are cultured on CIM, auxin signaling operates through *WOX11* and *LBD16*-mediated pathways, as well as through *PLT* and *CUC2*-mediated pathways, ultimately conferring pluripotency to callus cells (Fig. 2.1). During callus formation from explants, the transition of cell fate toward root meristem-like tissue is crucial for acquiring shoot regenerative competence. Auxin induces the expression of *PLT3*, *PLT5*, and *PLT7*, which subsequently activate *PLT1* and *PLT2*, genes characteristic of root meristem identity. Simultaneously, the shoot regeneration regulatory genes *CUC1* and *CUC2* are activated, enabling the healing tissues to acquire the capacity for shoot regeneration (Kareem et al. 2015). The induction of shoot regeneration requires the transfer of callus to a shoot-inducing medium (SIM). On SIM, callus tissues initially exhibit widespread auxin and cytokinin responses that gradually become spatially separated into distinct domains. Ultimately, the *WUS* gene is expressed in specific cells near the cytokinin response domain. Cytokinin signaling promotes shoot meristem formation through *ARR*-mediated activation of *WUS* expression. Additionally, the transcription factors PHABULOSA (*PHB*), PHAVOLUTA (*PHV*), and REVOLUTA (*REV*) contribute to shoot formation by inducing the expression of *WUS*, *STM*, and *RAP2.6 L* (Xu et al. 2024). Wounding can also stimulate shoot regeneration through the *WIND1*-dependent activation of *ESR1* (Fig. 2.1). The regeneration of roots from *Arabidopsis* leaf explants provides a simple and effective system for studying direct de novo root regeneration. In this model, root regeneration occurs without the addition of exogenous hormones and is entirely dependent on endogenous hormonal signaling within the leaf tissues. Following leaf excision, wound signals rapidly activate auxin biosynthesis genes *YUCCA5*, *YUCCA8*, and *YUCCA9*, leading to increased auxin production (Chen et al. 2016). The auxin response is strongly induced near the wound site, where *WOX11* and *WOX12* convert regeneration-competent vascular cells into root founder cells by activating the expression of *WOX5*, *WOX7*, and *LBD16*. These factors subsequently promote the establishment of root meristems and enhance regeneration efficiency. Under specific environmental stress conditions or hormonal induction, somatic cells can undergo somatic embryogenesis, either directly or indirectly. In direct somatic embryogenesis, somatic cells differentiate directly into embryos, whereas in indirect somatic embryogenesis, somatic cells first dedifferentiate into embryogenic callus before producing somatic embryos (Fehér 2015). Whether these two developmental pathways share identical molecular mechanisms remains unclear. When embryogenic callus is transferred to an auxin-free medium, polar auxin transport results in the formation of auxin maxima near the callus surface. During this process, cytokinin response domains initially overlap with auxin response domains. Auxin subsequently induces the expression of *WUS*, which is required for activating key embryonic regulators such as *LEC1* and LEC2. These regulators, together with *BBM* and *AGL15*, form a highly interconnected transcriptional network that promotes somatic embryogenesis through multiple positive feedback loops. Furthermore, these embryonic regulators stimulate the expression of *YUC*, *TAA1*, and IAA30, thereby modulating auxin biosynthesis and signaling pathways (Fig. 2.1) (Xu et al. 2024). Plant regeneration represents a complex biological process regulated by the coordinated interaction of multiple hormonal signaling pathways and transcriptional regulators. A comprehensive understanding of these molecular mechanisms is essential for elucidating the fundamental principles of plant regeneration and for providing theoretical and practical frameworks for the advancement of regeneration-based technologies in agriculture and plant biotechnology.

2.6 Conclusion

Developmental regulatory genes function as central molecular switches that govern the transition between differentiated and totipotent states in plants. Transcription factors such as *WUS/WOX*, *LEC*, *BBM*, *SERK*, *AGL15*, *PLT*, and *WIND* coordinate complex gene regulatory networks that integrate hormonal cues, stress signals, and epigenetic modifications to re-establish embryogenic competence. Their dynamic and environmentally dependent expression during tissue culture underlies cellular reprogramming, organogenesis, and somatic embryogenesis. Although ectopic expression of these individual regulators can enhance regeneration and transformation efficiency, totipotency is not controlled by a single master gene but arises from interconnected and species-specific regulatory pathways. The interaction between auxin-cytokinin signaling, chromatin remodeling, and transcriptional activation determines successful developmental switching. A better understanding of these coordinated molecular mechanisms will help in developing more effective regeneration systems and overcoming recalcitrance in economically important crops.

References

Abas L, Kolb M, Stadlmann J, Janacek DP, Lukic K, Schwechheimer C, Sazanov LA, Mach L, Friml J, Hammes UZ (2021) Naphthylphthalamic acid associates with and inhibits PIN auxin transporters. Proc Natl Acad Sci 118(1):e2020857118

Adamowski M, Friml J (2015) PIN-dependent auxin transport: action, regulation, and evolution. Plant Cell 27(1):20–32

Adero M, Tripathi JN, Tripathi L (2023) Advances in somatic embryogenesis of banana. Int J Mol Sci 24(13):10999

Agarwal Y, Shukla B, Manivannan A, Soundararajan P (2022) Paradigm and framework of WUS-CLV feedback loop in stem cell niche for SAM maintenance and cell identity transition. Agronomy 12(12):3132

Ai Z, Ma H, Zhang T, Chen S, Zhang J (2023) Molecular regulation of bud regeneration from callus of hybrid sweetgum (Liquidambar styracifluax Liquidambar formosana). Forests 14(9):1833

Bai J, Yang P, Xu L, Ming J (2025) LlLRP1, an SHI/SRS transcription factor, mediates bulbil formation in Lilium lancifolium via regulation by LlWOX11 and response to NaCl stress. Int J Biol Macromol 328:147385

Barreto HG, Ságio SA, Chalfun-Júnior A, Fevereiro P, Benedito VA (2019) Transcriptional profiling of the AFL subfamily of B3-type transcription factors during the in vitro induction of somatic embryogenesis in the model legume Medicago truncatula. Plant Cell Tissue Organ Cult 139(2):327–337

Berckmans B, Kirschner G, Gerlitz N, Stadler R, Simon R (2020) CLE40 signaling regulates root stem cell fate. Plant Physiol 182(4):1776–1792

Bidabadi SS, Jain SM (2020) Cellular, molecular, and physiological aspects of in vitro plant regeneration. Plants 9(6):702

Bilichak A, Luu J, Jiang F, Eudes F (2018) Identification of BABY BOOM homolog in bread wheat. Agri Gene 7:43–51

Bouchabke-Coussa O, Obellianne M, Linderme D, Montes E, Maia-Grondard A, Vilaine F, Pannetier C (2013) Wuschel overexpression promotes somatic embryogenesis and induces organogenesis in cotton (Gossypium hirsutum L.) tissues cultured in vitro. Plant Cell Rep 32(5):675–686

Bueno N, Cuesta C, Centeno ML, Ordás RJ, Alvarez JM (2021) In vitro plant regeneration in conifers: the role of WOX and KNOX gene families. Genes 12(3):438

Bull T (2022) Genetic analysis of in vitro regeneration and methods for enhancing regeneration and gene delivery in lettuce (Lactuca sativa L). Doctoral dissertation, University of California, Davis

Bull T, Michelmore R (2022) Molecular determinants of in vitro plant regeneration: prospects for enhanced manipulation of lettuce (Lactuca sativa L.). Front Plant Sci 13:888425

Bürglin TR, Affolter M (2016) Homeodomain proteins: an update. Chromosoma 125(3):497–521

Caeiro A, Caeiro S, Correia S, Canhoto J (2022) Induction of somatic embryogenesis in tamarillo (Solanum betaceum Cav.) involves increases in the endogenous auxin indole-3-acetic acid. Plants 11(10):1347

Cao H, Zhang X, Li F, Han Z, Ding B (2024) A point mutation in the IAA14 promoter enhances callus formation and regeneration. Physiol Mol Biol Plants 30(8):1253–1263

Chang Y, Liu J, Liu C, Liu H, Tang H, Qiu Y, Lin Z, Wang K, Yan Y, Ye X (2024) Establishment of a transformation system in close relatives of wheat under the assistance of TaWOX5. J Integr Agric 23(6):1839–1849

Chen L, Tong J, Xiao L, Ruan Y, Liu J, Zeng M, Huang H, Wang JW, Xu L (2016) YUCCA-mediated auxin biogenesis is required for cell fate transition occurring during de novo root organogenesis in Arabidopsis. J Exp Bot 67(14):4273–4284

Chen GZ, Huang J, Lin ZC, Wang F, Yang SM, Jiang X, Ahmad S, Zhou YZ, Lan S, Liu ZJ, Peng DH (2023) Genome-wide analysis of WUSCHEL-related homeobox gene family in sacred lotus (Nelumbo nucifera). Int J Mol Sci 24(18):14216

Chow V, Kirzinger MW, Kagale S (2023) Lend me your ears: a systematic review of the broad functions of ear motif-containing transcriptional repressors in plants. Genes 14(2):270

Conklin PA, Johnston R, Conlon BR, Shimizu R, Scanlon MJ (2020) Plant homeodomain proteins provide a mechanism for how leaves grow wide. Development 147(20):dev193623

Cueva-Agila AY, Alberca-Jaramillo N, Cella R, Concia L (2020) Isolation, phylogenetic analysis, and expression of a somatic embryogenesis receptor like kinase (SERK) gene in Cattleya maxima Lindl. Curr Plant Biol 21:100139

Dai X, Liu Z, Qiao M, Li J, Li S, Xiang F (2017) ARR12 promotes de novo shoot regeneration in Arabidopsis thaliana via activation of WUSCHEL expression. J Integr Plant Biol 59(10):747–758

De-la-Peña C (2013) New insights into somatic embryogenesis: LEAFY COTYLEDON1, BABY BOOM1 and WUSCHEL-RELATED HOMEOBOX4 are epigenetically regulated in Coffea canephora

Dhonukshe P, Tanaka H, Goh T, Ebine K, Mähönen AP, Prasad K, Blilou I, Geldner N, Xu J, Uemura T, Chory J (2008) Generation of cell polarity in plants links endocytosis, auxin distribution and cell fate decisions. Nature 456(7224):962–966

Dong X, Gao J, Jiang M, Tao Y, Chen X, Yang X, Wang L, Jiang D, Xiao Z, Bai X, He F (2024) The identification and characterization of WOX family genes in Coffea arabica reveals their potential roles in somatic embryogenesis and the cold-stress response. Int J Mol Sci 25(23):13031

Du F, Chang Z, Kong X, Xia L, Zhou W, Laux T, Zhang L (2025) Functional conservation and divergence of the WOX gene family in regulating meristem activity: from Arabidopsis to crops. Plant Physiol 199(1):kiaf374

Duan H, Maren NA, Ranney TG, Liu W (2022) New opportunities for using WUS/BBM and GRF-GIF genes to enhance genetic transformation of ornamental plants. Ornam Plant Res 2(1):1–7

El Ouakfaoui S, Schnell J, Abdeen A, Colville A, Labbé H, Han S, Baum B, Laberge S, Miki B (2010) Control of somatic embryogenesis and embryo development by AP2 transcription factors. Plant Mol Biol 74(4):313–326

Fambrini M, Durante C, Cionini G, Geri C, Giorgetti L, Michelotti V, Salvini M, Pugliesi C (2006) Characterization of LEAFY COTYLEDON1-LIKE gene in Helianthus annuus and its relationship with zygotic and somatic embryogenesis. Dev Genes Evol 216(5):253–264

Fambrini M, Usai G, Pugliesi C (2022) Induction of somatic embryogenesis in plants: different players and focus on WUSCHEL and WUS-RELATED HOMEOBOX (WOX) transcription factors. Int J Mol Sci 23(24):15950

Farina P, Fambrini M, Pugliesi C, Viviani A (2021) Expression of homeobox genes during in vitro culture of Lactuca sativa. Plant Biosyst Int J Deal Aspects Plant Biol 155(3):609–621

Fehér A (2015) Somatic embryogenesis—stress-induced remodeling of plant cell fate. Biochimica et Biophysica Acta (BBA)-Gene Regul Mech 1849(4):385–402

Fischer F, Grigolon G, Benner C, Ristow M (2022) Evolutionarily conserved transcription factors as regulators of longevity and targets for geroprotection. Physiol Rev 102:1449–1494

Florez SL, Erwin RL, Maximova SN, Guiltinan MJ, Curtis WR (2015) Enhanced somatic embryogenesis in Theobroma cacao using the homologous BABY BOOM transcription factor. BMC Plant Biol 15(1):121

Garcia C, Britto D, Marelli JP (2018) Transcription factors: their role in the regulation of somatic embryogenesis in Theobroma cacao L. and other species. Plant Cell Cult Protoc:385–396

Garcia C, Furtado de Almeida AA, Costa M, Britto D, Valle R, Royaert S, Marelli JP (2019) Abnormalities in somatic embryogenesis caused by 2, 4-D: an overview. Plant Cell Tissue Organ Cult 137(2):193–212

Guo J, Liu H, He Y, Cui X, Du X, Zhu J (2015) Origination of asexual plantlets in three species of Crassulaceae. Protoplasma 252(2):591–603

Han N, Tang R, Chen X, Xu Z, Ren Z, Wang L (2021) Genome-wide identification and characterization of WOX genes in Cucumis sativus. Genome 64(8):761–776

Hassani SB, Trontin JF, Raschke J, Zoglauer K, Rupps A (2022) Constitutive overexpression of a conifer WOX2 homolog affects somatic embryo development in Pinus pinaster and promotes somatic embryogenesis and organogenesis in Arabidopsis seedlings. Front Plant Sci 13:838421

He G, Cao Y, Wang J, Song M, Bi M, Tang Y, Xu L, Yang P, Ming J (2021) WUSCHEL-related homeobox genes cooperate with cytokinin signalling to promote bulbil formation in Lilium lancifolium. bioRxiv:2021–2010

Horstman A, Bemer M, Boutilier K (2017a) A transcriptional view on somatic embryogenesis. Regeneration 4(4):201–216

Horstman A, Li M, Heidmann I, Weemen M, Chen B, Muino JM, Angenent GC, Boutilier K (2017b) The BABY BOOM transcription factor activates the LEC1-ABI3-FUS3-LEC2 network to induce somatic embryogenesis. Plant Physiol 175(2):848–857

Hu H, Xiong L, Yang Y (2005) Rice SERK1 gene positively regulates somatic embryogenesis of cultured cell and host defense response against fungal infection. Planta 222(1):107–117

Hu B, Zhang G, Liu W, Shi J, Wang H, Qi M, Li J, Qin P, Ruan Y, Huang H, Zhang Y (2017) Divergent regeneration-competent cells adopt a common mechanism for callus initiation in angiosperms. Regeneration 4(3):132–139

Hu Y, Omary M, Hu Y, Doron O, Hoermayer L, Chen Q, Megides O, Chekli O, Ding Z, Friml J, Zhao Y (2021) Cell kinetics of auxin transport and activity in Arabidopsis root growth and skewing. Nat Commun 12(1):1657

Huang X, Sun MX (2025) Cell fate determination during sexual plant reproduction. New Phytol 245(2):480–495

Ikeuchi M, Sugimoto K, Iwase A (2013) Plant callus: mechanisms of induction and repression. Plant Cell 25(9):3159–3173

Ikeuchi M, Ogawa Y, Iwase A, Sugimoto K (2016) Plant regeneration: cellular origins and molecular mechanisms. Development 143(9):1442–1451

Ikeuchi M, Iwase A, Rymen B, Lambolez A, Kojima M, Takebayashi Y, Heyman J, Watanabe S, Seo M, De Veylder L, Sakakibara H (2017) Wounding triggers callus formation via dynamic hormonal and transcriptional changes. Plant Physiol 175(3):1158–1174

Ikeuchi M, Iwase A, Ito T, Tanaka H, Favero DS, Kawamura A, Sakamoto S, Wakazaki M, Tameshige T, Fujii H, Hashimoto N (2022) Wound-inducible WUSCHEL-RELATED HOMEOBOX 13 is required for callus growth and organ reconnection. Plant Physiol 188(1):425–441

Irikova T, Grozeva S, Denev I (2012) Identification of BABY BOOM and LEAFY COTYLEDON genes in sweet pepper (Capsicum annuum L.) genome by their partial gene sequences. Plant Growth Regul 67(2):191–198

Iwase A, Mita K, Nonaka S, Ikeuchi M, Koizuka C, Ohnuma M, Ezura H, Imamura J, Sugimoto K (2015) WIND1-based acquisition of regeneration competency in Arabidopsis and rapeseed. J Plant Res 128(3):389–397

Iwase A, Kondo Y, Laohavisit A, Takebayashi A, Ikeuchi M, Matsuoka K, Asahina M, Mitsuda N, Shirasu K, Fukuda H, Sugimoto K (2021) WIND transcription factors orchestrate wound-induced callus formation, vascular reconnection and defense response in Arabidopsis. New Phytol 232(2):734

Jha P, Ochatt SJ, Kumar V (2020) WUSCHEL: a master regulator in plant growth signaling. Plant Cell Rep 39(4):431–444

Jha S, Tyagi S, Choudhary D, Soni S, Singh J (2025) PIN-FORMED auxin efflux transporters in abiotic stress adaptation: mechanisms, expression dynamics, and crop improvement potential. Physiol Plant 177(5):e70556

Jiang Y, Wei X, Zhu M, Zhang X, Jiang Q, Wang Z, Cao Y, An X, Wan X (2024) Developmental regulators in promoting genetic transformation efficiency in maize and other plants. Curr Plant Biol 40:100383

Joshi S, Paul P, Hartman JM, Perry SE (2022) AGL15 promotion of somatic embryogenesis: role and molecular mechanism. Front Plant Sci 13:861556

Karami O, Rahimi A, Mak P, Horstman A, Boutilier K, Compier M, van der Zaal B, Offringa R (2021) An Arabidopsis AT-hook motif nuclear protein mediates somatic embryogenesis and coinciding genome duplication. Nat Commun 12(1):2508

Kareem A, Durgaprasad K, Sugimoto K, Du Y, Pulianmackal AJ, Trivedi ZB, Abhayadev PV, Pinon V, Meyerowitz EM, Scheres B, Prasad K (2015) PLETHORA genes control regeneration by a two-step mechanism. Curr Biol 25(8):1017–1030

Kerstens M, Galinha C, Hofhuis H, Nodine M, Pardal R, Scheres B, Willemsen V (2024) PLETHORA transcription factors promote early embryo development through induction of meristematic potential. Development 151(12):dev202527

Klimaszewska K, Pelletier G, Overton C, Stewart D, Rutledge RG (2010) Hormonally regulated overexpression of Arabidopsis WUS and conifer LEC1 (CHAP3A) in transgenic white spruce: implications for somatic embryo development and somatic seedling growth. Plant Cell Rep 29(7):723–734

Kowalik S, Samoń M, Przyborowski M (2026) Molecular regulators of in vitro regeneration in wheat: roles of morphogenic factors in transformation, genome editing, and breeding. Int J Mol Sci 27(3):1271

Kshetry AO, Ghose K, Alok A, Devkar V, Raman V, Stupar RM, Herrera-Estrella L, Zhang F, Patil GB (2025) Shoot at site: advancing in planta transformation, regeneration and gene-editing through a cascade of wounding-mediated developmental regulators. bioRxiv:2025–2002

Kurdyukov S, Song Y, Sheahan MB, Rose RJ (2014) Transcriptional regulation of early embryo development in the model legume Medicago truncatula. Plant Cell Rep 33(2):349–362

Kyo M, Maida K, Nishioka Y, Matsui K (2018) Coexpression of WUSCHEL related homeobox (WOX) 2 with WOX8 or WOX9 promotes regeneration from leaf segments and free cells in Nicotiana tabacum L. Plant Biotechnology 35(1):23–30

Li Z, Liu D, Xia Y, Li Z, Jing D, Du J, Niu N, Ma S, Wang J, Song Y, Yang Z (2020) Identification of the WUSCHEL-related homeobox (WOX) gene family, and interaction and functional analysis of TaWOX9 and TaWUS in wheat. Int J Mol Sci 21(5):1581

Li M, Wrobel-Marek J, Heidmann I, Horstman A, Chen B, Reis R, Angenent GC, Boutilier K (2022) Auxin biosynthesis maintains embryo identity and growth during BABY BOOM-induced somatic embryogenesis. Plant Physiol 188(2):1095–1110

Li Y, Yu J, Miao J, Yue W, Xu T (2025) Identification of the embryogenesis gene BBM in alfalfa (Medicago sativa) and analysis of its expression pattern. Agronomy 15(8):1768

Lian Z, Nguyen CD, Liu L, Wang G, Chen J, Wang S, Yi G, Wilson S, Ozias-Akins P, Gong H, Huo H (2022) Application of developmental regulators to improve in planta or in vitro transformation in plants. Plant Biotechnol J 20(8):1622–1635

Liu JingChun, L. J., Sheng LiHong, S. L., Xu YingQiang, X. Y., Li JiQin, L. J., Yang ZhongNan, Y. Z., Huang Hai, H. H., & Xu Lin, X. L. (2014). WOX11 and 12 are involved in the first-step cell fate transition during de novo root organogenesis in Arabidopsis

Liu X, Bie XM, Lin X, Li M, Wang H, Zhang X, Yang Y, Zhang C, Zhang XS, Xiao J (2023) Uncovering the transcriptional regulatory network involved in boosting wheat regeneration and transformation. Nat Plants 9(6):908–925

Long Y, Yang Y, Pan G, Shen Y (2022) New insights into tissue culture plant-regeneration mechanisms. Front Plant Sci 13:926752

Long X, Zhang J, Wang D, Weng Y, Liu S, Li M, Hao Z, Cheng T, Shi J, Chen J (2023) Expression dynamics of WOX homeodomain transcription factors during somatic embryogenesis in Liriodendron hybrids. For Res 3:15

Lou H, Huang Y, Wang W, Cai Z, Cai H, Liu Z, Sun L, Xu Q (2022) Overexpression of the AtWUSCHEL gene promotes somatic embryogenesis and lateral branch formation in birch (Betula platyphylla Suk.). Plant Cell Tissue Organ Cult 150(2):371–383

Luo Z, Wu L, Miao X, Zhang S, Wei N, Zhao S, Shang X, Hu H, Xue J, Zhang T, Yang F (2024) A dynamic regulome of shoot-apical-meristem-related homeobox transcription factors modulates plant architecture in maize. Genome Biol 25(1):245

Ma J, Wang Y, Chen S, Xu T (2025) The central role of auxin in orchestrating apical stem cells in plants. Plant Cell Environ 48:5053–5072

Malabadi RB, Chalannavar RK, Kolkar KP (2025) WUCHEL gene family: transcription factor-WOX2 as a early genetic marker of gene expression during induction of somatic embryogenesis: An updated review. World J Adv Eng Technol Sci 16(03):552–582

Mehbub H, Akter A, Akter MA, Mandal MSH, Hoque MA, Tuleja M, Mehraj H (2022) Tissue culture in ornamentals: cultivation factors, propagation techniques, and its application. Plants 11(23):3208

Mehraj H, Maruyama T, Tsuruta M, Ueno S (2025) Developmental regulatory genes: key switches in the developmental events of plant tissue culture. Curr Plant Biol 44:100550

Méndez-Hernández HA, Ledezma-Rodríguez M, Avilez-Montalvo RN, Juárez-Gómez YL, Skeete A, Avilez-Montalvo J, De-la-Peña C, Loyola-Vargas VM (2019) Signaling overview of plant somatic embryogenesis. Front Plant Sci 10:77

Mookkan M, Nelson-Vasilchik K, Hague J, Zhang ZJ, Kausch AP (2017) Selectable marker independent transformation of recalcitrant maize inbred B73 and sorghum P898012 mediated by morphogenic regulators BABY BOOM and WUSCHEL2. Plant Cell Rep 36(9):1477–1491

Orłowska A, Kępczyńska E (2018) Identification of Polycomb repressive Complex1, Trithorax group genes and their simultaneous expression with WUSCHEL, WUSCHEL-related Homeobox5 and SHOOT MERISTEMLESS during the induction phase of somatic embryogenesis in Medicago truncatula Gaertn. Plant Cell Tissue Organ Cult 134(3):345–356

Orłowska A, Igielski R, Łagowska K, Kępczyńska E (2017) Identification of LEC1, L1L and Polycomb Repressive Complex 2 genes and their expression during the induction phase of Medicago truncatula Gaertn. somatic embryogenesis. Plant Cell Tissue Organ Cult 129(1):119–132

Pan J, Zhao F, Zhang G, Pan Y, Sun L, Bao N, Qin P, Chen L, Yu J, Zhang Y, Xu L (2019) Control of de novo root regeneration efficiency by developmental status of Arabidopsis leaf explants. J Genet Genomics 46(3):133–140

Paul P, Joshi S, Tian R, Diogo Junior R, Chakrabarti M, Perry SE (2022) The MADS-domain factor AGAMOUS-Like18 promotes somatic embryogenesis. Plant Physiol 188(3):1617–1631

Phong TH, Hieu T, Tung HT, Mai NTN, Khai HD, Cuong DM, Luan VQ, Nam NB, Nhut DT (2023) Somatic embryogenesis as potential method for commercial propagation in Passiflora edulis Sims f. edulis-an important horticultural crop. Sci Hortic 316:112020

Pokimica N, Ćosić T, Uzelac B, Ninković S, Raspor M (2024) Dissecting the roles of the cytokinin signaling network: the case of de novo shoot apical meristem formation. Biomolecules 14(3):381

Qi S, Zhao R, Yan J, Fan Y, Huang C, Li H, Chen S, Zhang T, Kong L, Zhao J, Zhang J (2021) Global transcriptome and coexpression network analyses reveal new insights into somatic embryogenesis in hybrid sweetgum (Liquidambar styracifluax Liquidambar formosana). Front Plant Sci 12:751866

Rashid SZ, Yamaji N, Kyo M (2007) Shoot formation from root tip region: a developmental alteration by WUS in transgenic tobacco. Plant Cell Rep 26(9):1449–1455

Ribeiro C, de Melo BP, Lourenço-Tessutti IT, Ballesteros HF, Ribeiro KVG, Menuet K, Heyman J, Hemerly A, de Sá MFG, De Veylder L, de Almeida Engler J (2024) The regeneration conferring transcription factor complex ERF115-PAT1 coordinates a wound-induced response in root-knot nematode induced galls. New Phytol 241(2):878–895

Riccucci E, Vanni C, Vangelisti A, Fambrini M, Giordani T, Cavallini A, Mascagni F, Pugliesi C (2023) Genome-wide analysis of WOX multigene family in sunflower (Helianthus annuus L.). Int J Mol Sci 24(4):3352

Rodríguez-Herrera J, Galván-Alcaraz KA, Albarrán-Hernández RU, Villa-Núñez JP, Rodríguez-Alonso G, Shishkova S (2025) A computational Evo-Devo approach for elucidating the roles of PLETHORA transcription factors in regulating root development. PLoS One 20(7):e0327511

Salaün C, Lepiniec L, Dubreucq B (2021) Genetic and molecular control of somatic embryogenesis. Plants 10(7):1467. https://doi.org/10.3390/plants10071467

Sarkar AK, Luijten M, Miyashima S, Lenhard M, Hashimoto T, Nakajima K, Scheres B, Heidstra R, Laux T (2007) Conserved factors regulate signalling in Arabidopsis thaliana shoot and root stem cell organizers. Nature 446(7137):811–814

Schaller A (2024) REGENERATION FACTOR 1, a peptide boost for wound healing and plant biotechnology. Mol Plant 17(9):1333–1334

Sengupta S, Nag Chaudhuri R (2020) ABI3 plays a role in de-novo root regeneration from Arabidopsis thaliana callus cells. Plant Signal Behav 15(9):1794147

Shah K, Schmidt ED, Vlak JM, de Vries SC (2001) Expression of the Daucus carota somatic embryogenesis receptor kinase (DcSERK) protein in insect cells. Biochimie 83(5):415–421

Shires ME, Florez SL, Lai TS, Curtis WR (2023) Editorial expression of concern: inducible somatic embryogenesis in Theobroma cacao achieved using the DEX-activatable transcription factor-glucocorticoid receptor fusion. Biotechnol Lett 45(10):1393–1393

Šmeringai J, Schrumpfová PP, Pernisová M (2023) Cytokinins–regulators of de novo shoot organogenesis. Front Plant Sci 14:1239133

Su YH, Tang LP, Zhao XY, Zhang XS (2021) Plant cell totipotency: insights into cellular reprogramming. J Integr Plant Biol 63(1):228–243

Suzuki M, Yamazaki C, Mitsui M, Kakei Y, Mitani Y, Nakamura A, Ishii T, Soeno K, Shimada Y (2015) Transcriptional feedback regulation of YUCCA genes in response to auxin levels in Arabidopsis. Plant Cell Rep 34(8):1343–1352

Tang LP, Zhou C, Wang SS, Yuan J, Zhang XS, Su YH (2017) FUSCA 3 interacting with LEAFY COTYLEDON 2 controls lateral root formation through regulating YUCCA 4 gene expression in Arabidopsis thaliana. New Phytol 213(4):1740–1754

Tang Y, Wu W, Zheng X, Lu L, Chen X, Hao Z, Liu S, Chen Y (2023) AT-hook transcription factors show functions in Liriodendron chinense under drought stress and somatic embryogenesis. Plants 12(6):1353

Tian R, Wang F, Zheng Q, Niza VM, Downie AB, Perry SE (2020) Direct and indirect targets of the arabidopsis seed transcription factor ABSCISIC ACID INSENSITIVE3. Plant J 103(5):1679–1694

Tokuji Y, Takano S, Tonomura M, Tanaka S, Igari K, Watanabe T (2011) Influence of 5'-azacitidine on promoting recovery of cell competence for shoot organogenesis in Arabidopsis. Plant Cell Tissue Organ Cult 106(2):289–297

Torres-Silva G, Correia LNDF, Koehler AD, Batista DS, Faria DV, Resende SV, Strickler SR, Fouracre J, Romanel E, Specht CD, Otoni WC (2021) Expression of Melocactus glaucescens SERK1 sheds new light on the mechanism of areolar activation in cacti. Plant Cell Tissue Organ Cult 147(3):437–451

Tvorogova VE, Krasnoperova EY, Potsenkovskaia EA, Kudriashov AA, Dodueva IE, Lutova LA (2021) What does the WOX say? Review of regulators, targets, partners. Mol Biol 55(3):311–337

van der Graaff E, Laux T, Rensing SA (2009) The WUS homeobox-containing (WOX) protein family. Genome Biol 10(12):248

Verma S, Attuluri VPS, Robert HS (2021) An essential function for auxin in embryo development. Cold Spring Harb Perspect Biol 13(4):a039966

Vollbrecht E, Veit B, Sinha N, Hake S (1991) The developmental gene Knotted-1 is a member of a maize homeobox gene family. Nature 350(6315):241–243

Wang Y, Jiao Y (2023) Cell signaling in the shoot apical meristem. Plant Physiol 193(1):70–82

Wang FX, Shang GD, Wu LY, Xu ZG, Zhao XY, Wang JW (2020) Chromatin accessibility dynamics and a hierarchical transcriptional regulatory network structure for plant somatic embryogenesis. Dev Cell 54(6):742–757

Wang P, Si H, Li C, Xu Z, Guo H, Jin S, Cheng H (2025) Plant genetic transformation: achievements, current status and future prospects. Plant Biotechnol J 23(6):2034–2058

Winnicki K (2020) The winner takes it all: auxin—the main player during plant embryogenesis. Cells 9(3):606

Wójcik AM, Wójcikowska B, Gaj MD (2020) Current perspectives on the auxin-mediated genetic network that controls the induction of somatic embryogenesis in plants. Int J Mol Sci 21(4):1333

Wójcik AM, Krypczyk K, Buchcik WM, Gaj MD (2025) From root to embryogenic transition: WOX5 reprograms plant somatic cells via auxin-mediated pathways. BMC Plant Biol 25(1):642

Wójcikowska B, Wójcik AM, Gaj MD (2020) Epigenetic regulation of auxin-induced somatic embryogenesis in plants. Int J Mol Sci 21(7):2307

Wu H, Qu X, Dong Z, Luo L, Shao C, Forner J, Lohmann JU, Su M, Xu M, Liu X, Zhu L (2020) WUSCHEL triggers innate antiviral immunity in plant stem cells. Science 370(6513):227–231

Xu C, Cao H, Zhang Q, Wang H, Xin W, Xu E, Zhang S, Yu R, Yu D, Hu Y (2018) Control of auxin-induced callus formation by bZIP59–LBD complex in Arabidopsis regeneration. Nat Plants 4(2):108–115

Xu P, Zhong Y, Xu A, Liu B, Zhang Y, Zhao A, Yang X, Ming M, Cao F, Fu F (2024) Application of developmental regulators for enhancing plant regeneration and genetic transformation. Plants 13(9):1272

Xu X, He X, Zhang Q, Yang L (2025) Genome-wide identification and expression pattern analysis of nuclear factor YB/C genes in Pinus koraiensis, and functional identification of Leafy cotyledon 1. Plants 14(3):438

Xue Z, Liu L, Zhang C (2020) Regulation of shoot apical meristem and axillary meristem development in plants. Int J Mol Sci 21(8):2917

Yamoune A, Cuyacot AR, Zdarska M, Hejatko J (2021) Hormonal orchestration of root apical meristem formation and maintenance in Arabidopsis. J Exp Bot 72(19):6768–6788

Yang HF, Kou YP, Gao B, Soliman TMA, Xu KD, Ma N, Cao X, Zhao LJ (2014) Identification and functional analysis of BABY BOOM genes from Rosa canina. Biol Plant 58(3):427–435

Yang Q, Yuan C, Cong T, Zhang Q (2023) The secrets of meristems initiation: axillary meristem initiation and floral meristem initiation. Plants 12(9):1879

Yavuz C, Tek AL, Çalışkan ME (2024) Induction of chromosomal aberrations with ectopic expression of native BABY BOOM1 (GmBBM1) in soybean. Plant Cell Tissue Organ Cult 158(3):59

Yuan HY, Kagale S, Ferrie AM (2024) Multifaceted roles of transcription factors during plant embryogenesis. Front Plant Sci 14:1322728

Zhai N, Xu L (2021) Pluripotency acquisition in the middle cell layer of callus is required for organ regeneration. Nat Plants 7(11):1453–1460

Zhang S, Wong L, Meng L, Lemaux PG (2002) Similarity of expression patterns of knotted1 and ZmLEC1 during somatic and zygotic embryogenesis in maize (Zea mays L.). Planta 215(2):191–194

Zhang K, Su J, Xu M, Zhou Z, Zhu X, Ma X, Hou J, Tan L, Zhu Z, Cai H, Liu F (2020a) A common wild rice-derived BOC1 allele reduces callus browning in indica rice transformation. Nat Commun 11(1):443

Zhang T, Li C, Li D, Liu Y, Yang X (2020b) Roles of YABBY transcription factors in the modulation of morphogenesis, development, and phytohormone and stress responses in plants. J Plant Res 133(6):751–763

Zhang Z, Conner J, Guo Y, Ozias-Akins P (2020c) Haploidy in tobacco induced by PsASGR-BBML transgenes via parthenogenesis. Genes 11(9):1072

Zhang ZA, Liu MY, Ren SN, Liu X, Gao YH, Zhu CY, Niu HQ, Chen BW, Liu C, Yin W, Wang HL (2023) Identification of WUSCHEL-related homeobox gene and truncated small peptides in transformation efficiency improvement in eucalyptus. BMC Plant Biol 23(1):604

Zhao Y, Cheng S, Song Y, Huang Y, Zhou S, Liu X, Zhou DX (2015) The interaction between rice ERF3 and WOX11 promotes crown root development by regulating gene expression involved in cytokinin signaling. Plant Cell 27(9):2469–2483

Zheng Q, Zheng Y, Ji H, Burnie W, Perry SE (2016) Gene regulation by the AGL15 transcription factor reveals hormone interactions in somatic embryogenesis. Plant Physiol 172(4):2374–2387

Zhou X, Guo Y, Zhao P, Sun MX (2018) Comparative analysis of WUSCHEL-related homeobox genes revealed their parent-of-origin and cell type-specific expression pattern during early embryogenesis in tobacco. Front Plant Sci 9:311

Zhou R, Zhao Y, Cheng P, Zhang B, Liu Z, Wang S, Li H, Chen Q, Zhao Y, Li S, Wu X (2023) GmBBM7 promotes callus and root growth during somatic embryogenesis of soybean (Glycine max). Biotechnol Biotechnol Equip 37(1):2238833

Zhu T, Moschou PN, Alvarez JM, Sohlberg JJ, von Arnold S (2016) WUSCHEL-RELATED HOMEOBOX 2 is important for protoderm and suspensor development in the gymnosperm Norway spruce. BMC Plant Biol 16(1):19

Zhu Y, Xie L, Chen GQ, Lee MY, Loque D, Scheller HV (2018) A transgene design for enhancing oil content in Arabidopsis and Camelina seeds. Biotechnol Biofuels 11(1):46

Zuo J, Niu QW, Frugis G, Chua NH (2002) The WUSCHEL gene promotes vegetative-to-embryonic transition in Arabidopsis. Plant J 30(3):349–359

Chapter 3
Hormone-Responsive Gene Modules in Regeneration

Abstract Plant regeneration is one of the most remarkable features of plant development. Unlike most animals, plants can reprogram fully differentiated somatic cells and redirect them toward the formation of new tissues, organs, or whole plants. This developmental flexibility forms the foundation of plant tissue culture and reinforces modern technologies such as micropropagation, genetic transformation, and genome editing. Auxin and cytokinin work together to regulate hormone-responsive gene networks, which are at the core of this process. For decades, regeneration was explained by the classical hormone ratio concept, where auxin promotes root formation and callus induction, while cytokinin drives shoot development. Although this theory remains important, recent molecular advances show that regeneration is significantly more complex than a simple hormonal balance. Auxin is detected by *TRANSPORT INHIBITOR RESPONSE 1/AUXIN SIGNALING F-BOX* (*TIR1/AFB*) receptor complexes, leading to activation of *AUXIN RESPONSE FACTORS* (*ARF*) that stimulate genes such as *LATERAL ORGAN BOUNDARIES DOMAIN* (*LBD*), resulting in cellular reprogramming. In parallel, cytokinin signaling operates through a multistep phosphorelay pathway involving histidine kinase receptors and *AUXIN RESPONSE REGULATORS* (ARRs), ultimately activating meristem regulators such as *WUSCHEL* (*WUS*) to establish shoot apical meristem (*SAM*) identity. Importantly, auxin and cytokinin pathways do not function independently. They intersect at several regulatory sites, including hormone biosynthesis, transport, and transcriptional regulation, to build feedback-driven gene modules that guide cell fate decisions. Epigenetic processes further shape this process by altering chromatin accessibility and hormone sensitivity. Comparative studies in model plants such as *Arabidopsis thaliana*, *Oryza sativa*, and *Zea mays* reveal a common hormonal framework while also highlighting species-specific constraints, particularly in cereals where genotype dependence and epigenetic barriers can limit regeneration efficiency.

Keywords Acquisition of competence · Auxin-cytokinin signaling · Dedifferentiation · Hormone crosstalk · Plant regeneration · Redifferentiation

P. Kumar, *Totipotency in Plants*, SpringerBriefs in Plant Science,
https://doi.org/10.1007/978-981-92-1631-4_3

3.1 Introduction

Organisms often regenerate lost or damaged parts to reduce the impact of injury, and this ability to restore tissues from simple wound repair to the emergence of new organs or whole individuals varies greatly across different groups of animals and plants. Plants possess greater regenerative plasticity, wherein fully differentiated somatic cells are capable of reverting to a pluripotent or totipotent state and subsequently giving rise to entire plants. This remarkable capacity forms the biological foundation of plant tissue culture technologies and has been utilized in clonal propagation practices such as cuttings and grafting and is now being established at molecular and cellular levels in recent studies. Strategies for regenerating whole plants from small tissues or single cells in vitro began in the early twentieth century with Haberlandt's vision toward the plant tissue culture concept. In vitro plant regeneration is a process in which explants, after undergoing cell division and differentiation, form organs and tissues throughout their growth period. This process strengthens modern plant biotechnology, including micropropagation, genetic transformation, and genome editing (Bidabadi and Jain 2020). Regeneration proceeds through a coordinated sequence of dedifferentiation, acquisition of competence, and redifferentiation, all of which are tightly regulated by plant hormones and their downstream gene regulatory networks (Kutschera and Niklas 2013). Among the various endogenous and exogenous applied signals, the plant hormones such as auxin and cytokinin play a central and antagonistic role in controlling in vitro regeneration. Subsequent molecular studies have revealed that this hormonal balance operates through precise regulation of gene expression, signal transduction pathway, epigenetic modification, determining cell fate by regulating transcriptional programs, and developmental regulator (DR) gene expression (Sugimoto et al. 2019). The hormone auxin generally favors root formation, whereas cytokinin promotes shoot regeneration. This basic hormonal idea continues to serve as the foundation for modern plant regeneration studies and has been enhanced by subsequent molecular and genetic discoveries into hormone-regulated cell fate transitions (Skoog 1957). This chapter focuses on the molecular interaction between the auxin and cytokinin signaling and gene regulation during in vitro regeneration, highlighting the hormone gradients, signal transduction components, transcriptional reprogramming, and feedback regulation. The simplistic hormone-ratio model has evolved into a sophisticated understanding of hormone gradients, signaling networks, and transcriptional regulation. Advances in molecular genetics, especially in *Arabidopsis thaliana*, revealed that hormones act through defined receptors, signal transduction cascades, and transcription factor networks (Xu et al. 2024). Regeneration is now understood as a multistep process involving dedifferentiation, acquisition of competence, and redifferentiation, each governed by precise hormonal and genetic controls. This chapter integrates classical concepts with modern molecular insights, emphasizing how auxin-cytokinin interplay regulates gene expression and cell fate during in vitro regeneration. Case studies from *Arabidopsis thaliana*, *Zea mays* (maize), and *Oryza sativa* (rice) have been discussed to illustrate species-specific mechanisms underlying hormone-driven regeneration.

3.2 Cellular Phases of In Vitro Regeneration

Regeneration typically proceeds through three interconnected cellular phases, namely, dedifferentiation, acquisition of competence, and redifferentiation. Dedifferentiation involves the loss of the original cellular identity and re-entry into the cell cycle; acquisition of competence refers to the establishment of a hormone-responsive and developmentally plastic state; and redifferentiation entails the commitment of competent cells to specific organ identities (Fig. 3.1) (Ikeuchi et al. 2016).

3.2.1 *Dedifferentiation*

Dedifferentiation marks the initial phase of in vitro regeneration in which differentiated somatic cells lose their specialized functions and original identities, re-enter the cell cycle, proliferate and regenerate tissues and organs, and even become complete plants. Morphologically, dedifferentiation is characterized by cell enlargement, an increased nuclear-to-cytoplasmic ratio, and reactivation of cell division. At the molecular level, dedifferentiation involves widespread repression of tissue-specific genes and activation of genes associated with cell cycle progression, chromatin remodeling, and hormone responsiveness. Several reports have shown the totipotent ability of plant cells through which plants can be regenerated, which in turn is widely used in several basic studies such as in micropropagation, germplasm conservation, and formation of genetically modified plants (Sugimoto et al. 2019). This differentiation process is predominantly triggered by auxin accumulation.

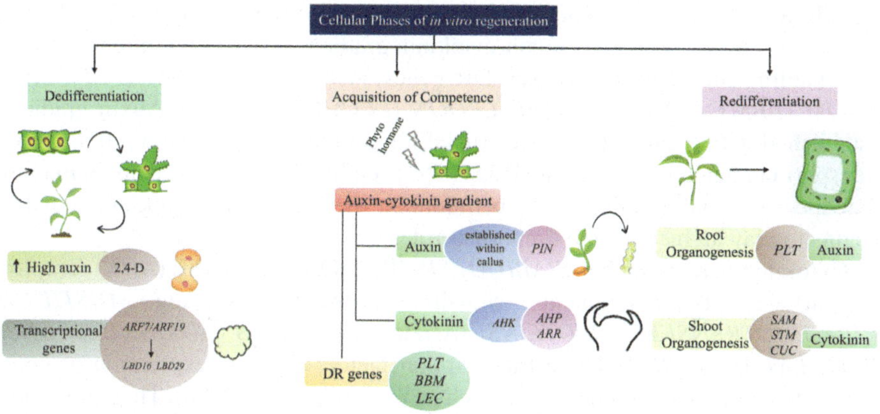

Fig. 3.1 Cellular phases of in vitro regeneration (*AHK*, Arabidopsis Histidine Kinase, *AHP* Histidine Phosphotransfer Proteins, *ARR* Auxin Response Regulator, *ARF7* Auxin Response Factor 7, *ARF19* Auxin Response Factor 19, *BBM* Baby Boom, *CUC* Cup-Shaped Cotyledon, *LBD16/29* Lateral Organ Boundaries Domain 16/29, *LEC* Leafy Cotyledon, *PIN* Pin-Formed, *PLT* Plethora, *SAM* Shoot Apical Meristem, *STM* Shoot Meristemless, 2,4-dichlorophenoxyacetic acid)

High auxin concentrations in callus induction media promote the accumulation of auxin maxima within explant tissues through enhanced biosynthesis and polar transport. High exogenous auxin concentrations, particularly synthetic auxins such as 2,4-dichlorophenoxyacetic acid (2,4-D), induce cell division and promote callus formation. Auxin perception occurs via *TIR1/AFB* receptors, which mediate the degradation of *AUXIN/INDOLE-3-ACETIC ACID* (*Aux/IAA*) repressor proteins. This degradation releases ARFs, enabling them to activate transcription of auxin-responsive genes (Leyser 2018). A key transcriptional module activated during dedifferentiation is the *ARF7/ARF19-LBD* pathway. *ARF7* and *ARF19* directly induce *LBD* genes such as *LBD16* and *LBD29*, which promote callus formation by regulating cell division and chromatin accessibility (Vissenberg et al. 2020). Genetic studies in *Arabidopsis* have demonstrated that loss-of-function mutations in *ARF7*, *ARF19*, or *LBD* genes severely impair callus initiation, underscoring their central role in dedifferentiation. Notably, auxin-induced callus often exhibits molecular signatures resembling root meristem identity, even when derived from non-root tissues, highlighting the conserved nature of auxin-mediated dedifferentiation programs (Banda et al. 2019).

3.2.2 Acquisition of Competence

Dedifferentiated cells enter the transitional state called the acquisition of competence. In this state the dedifferentiated cells become responsive to developmental cues that determine organogenic fate. Competence is not an inherent property of all dedifferentiated cells but is established through sustained hormone signaling, transcriptional stabilization, and epigenetic remodeling. During this phase, auxin and cytokinin gradients are refined, and cells acquire differential sensitivity to these hormones. During this stage, auxin-cytokinin gradients establish hormone sensitivity, enabling the activation of key DR genes such as *PLETHORA* (*PLT*), *BABY BOOM* (*BBM*), and *LEAFY COTYLEDON* (*LEC*) (Fehér 2019). Auxin gradients established within callus tissue are crucial for competence acquisition. Polar auxin transport mediated by PIN-FORMED (*PIN*) proteins generates local auxin maxima that specify founder cells for regeneration. At the same time, cytokinin signaling begins to play an increasingly important role.

Cytokinin perception by *ARABIDOPSIS* HISTIDINE KINASE (*AHK*) receptors activate a two-component signaling system involving *HISTIDINE PHOSPHOTRANSFER PROTEINS* (*AHPs*) and type-B *AUXIN RESPONSE REGULATORS* (*ARRs*). These transcription factors promote expression of genes associated with meristem initiation, such as *WUS* (Che et al. 2006). These hormone-responsive gene modules are stabilized by epigenetic remodeling and signal transduction components, preparing cells for organ-specific differentiation. Epigenetic regulation plays a crucial role during this phase. Dedifferentiation and competence acquisition are accompanied by changes in histone modifications, such as reduced levels of the repressive mark *H3K27me3* (trimethylation of lysine 27 on histone H3) and increased histone acetylation at regeneration-associated loci. DNA methylation patterns are also dynamically altered, contributing to the stabilization of competence-related gene expression. These epigenetic changes ensure that regeneration-competent cells maintain a transcriptional state favorable for redifferentiation (Li et al. 2024).

3.2.3 Redifferentiation

Redifferentiation is the final phase in which competent cells commit to specific organ identities. Depending on culture conditions, redifferentiation can result in shoot organogenesis, root organogenesis, or somatic embryogenesis. Cytokinin is the primary driver of shoot redifferentiation. Elevated cytokinin signaling activates type-B ARRs and promotes shoot regeneration through the activation of *WUS* expression in localized domains, leading to the formation of *SHOOT APICAL MERISTEM* (*SAM*), *SHOOT MERISTEMLESS* (*STM*), and *CUP-SHAPED COTYLEDON* (*CUC*) genes, whereas auxin-dominant conditions favor root formation. Feedback regulation between hormone signaling pathways and DR gene expression ensures spatial and temporal precision during organogenesis (Kieber and Schaller 2018). Root redifferentiation, by contrast, is associated with sustained auxin signaling and activation of *PLT* transcription factors. These factors establish root stem cell niches and regulate pattern formation. The balance between auxin and cytokinin thus determines the developmental trajectory of regenerating tissues (Su et al. 2011).

3.3 Auxin Signaling and Transcriptional Control

Auxin signaling operates through a highly conserved and well-characterized molecular pathway that translates hormone perception into transcriptional responses essential for plant regeneration. Auxin is perceived by the *TIR1/AFB* family of receptors, which function as part of the SCF ubiquitin ligase complex. Upon auxin binding, *TIR1/AFB* receptors promote the ubiquitination and subsequent proteasomal degradation of *AUX/IAA* transcriptional repressors. This degradation releases *ARF* transcription factors, enabling them to activate or repress downstream auxin-responsive genes (Kepinski and Leyser 2005). During in vitro regeneration, specific *ARFs* play pivotal regulatory roles. *ARF5*, also known as *MONOPTEROS* (*MP*), is a master regulator of embryonic patterning and vascular development and has been shown to be essential for establishing auxin-responsive developmental axes during regeneration. *ARF7* and *ARF19* function redundantly and are particularly important for dedifferentiation and callus initiation. These *ARF*s directly activate *LBD* transcription factors, such as *LBD16*, *LBD18*, and *LBD29*, which promote cell cycle re-entry and proliferation, thereby driving the transition of differentiated cells into a pluripotent state (Fan et al. 2012). Auxin signaling during regeneration has been further characterized by the formation of self-reinforcing regulatory loops. *ARFs* not only regulate genes involved in cell division and pattern formation but also control the expression of auxin biosynthesis and transport genes, thereby stabilizing local auxin maxima. This feedback ensures sustained auxin signaling at regeneration sites, a process crucial for maintaining developmental competence (De Rybel et al. 2014). Spatial and temporal specificity of auxin signaling has been achieved through the coordinated action of polar auxin transport and localized auxin biosynthesis. Polar auxin transport is mediated primarily by *PIN* efflux carriers, whose asymmetric localization establishes directional auxin fluxes and concentration gradients within regenerating tissues. These gradients act as positional cues that specify founder cells for organ initiation and pattern formation (Petrášek and Friml 2009). In addition to transport, local auxin biosynthesis plays a critical role in regeneration. The *YUCCA* (*YUC*) family of genes encodes flavin monooxygenases that catalyze a rate-limiting step in the tryptophan-dependent auxin biosynthesis pathway. Several *YUC* genes are strongly upregulated during callus formation and organ regeneration, leading to localized auxin accumulation independent of long-distance transport. This localized biosynthesis ensures a continuous auxin supply at regeneration sites, reinforcing auxin signaling domains and enabling robust developmental reprogramming (Sugawara et al. 2009). Collectively, the integration of auxin perception, *ARF*-mediated transcriptional control, polar transport, and local biosynthesis forms a dynamic and self-organizing regulatory network. This network underpins the plasticity of plant cells during in vitro regeneration, allowing precise control of cell fate transitions from dedifferentiation to organized organ formation.

3.4 Cytokinin Signaling and Shoot Regeneration

Cytokinin signaling is mediated through a conserved multistep phosphorelay system that closely resembles bacterial two-component signaling pathways. Cytokinin perception occurs at the plasma membrane and endoplasmic reticulum through *AHK* receptors, primarily *AHK2*, *AHK3*, and *AHK4/CRE1*. Upon cytokinin binding, these receptors undergo autophosphorylation at a conserved histidine residue, initiating a phosphorylation cascade that transmits the signal through AHPs to downstream response regulators in the nucleus (Hwang et al. 2012). The final output of cytokinin signaling is mediated by ARRs, which are classified into type-A and type-B based on their structure and function. Type-B ARRs are transcription factors containing a receiver domain and a C-terminal *MYB*-like DNA-binding domain. Upon phosphorylation, type-B ARRs bind to cytokinin-responsive promoter elements and activate transcription of genes involved in cell division, meristem maintenance, and shoot identity (Argyros et al. 2008).

During in vitro shoot regeneration, cytokinin signaling plays a decisive role in establishing a functional *SAM*. Elevated cytokinin levels activate type-B *ARRs* such as *ARR1*, *ARR10*, and *ARR12*, which directly induce the expression of key meristem regulators including *WUS* and *STM*. *WUS* is essential for specifying and maintaining stem cell identity within the organizing center of the *SAM*, while *STM* prevents premature differentiation of meristematic cells by repressing differentiation pathways. Together, *WUS* and *STM* establish a self-sustaining meristematic domain that supports continuous shoot formation (Gordon et al. 2009; Su et al. 2011). Cytokinin-mediated induction of *WUS* during regeneration occurs in a spatially restricted manner, typically in localized regions of callus tissue exposed to shoot-inducing medium. This spatial precision is critical for proper shoot patterning and has been achieved through tight regulation of cytokinin signal intensity and duration. Excessive or ectopic cytokinin signaling can lead to abnormal meristem formation or uncontrolled shoot proliferation, underscoring the importance of regulatory feedback mechanisms. Negative feedback regulation within the cytokinin signaling pathway is primarily mediated by type-A *ARRs*, such as *ARR5*, *ARR6*, *ARR7*, and *ARR15*. These proteins lack a DNA-binding domain and function mainly as negative regulators by competing with type-B *ARRs* for phosphorylation or interacting with signaling components to attenuate signal transduction. Importantly, type-A *ARR* genes are themselves rapidly induced by cytokinin, forming a classic negative feedback loop that dampens cytokinin responses and stabilizes meristem development (To et al. 2007; Kieber and Schaller 2018). This feedback regulation ensures precise spatial and temporal control of shoot initiation during regeneration. By limiting the extent of cytokinin signaling, type-A *ARRs* prevent ectopic activation of meristem genes outside the intended regeneration zones, thereby maintaining organized shoot formation. The balance between type-B *ARR*-mediated activation and type-A *ARR*-mediated repression allows regenerating tissues to fine-tune cytokinin responses according to developmental context and environmental conditions. Overall, cytokinin signaling integrates hormone perception, transcriptional regulation, and feedback control to orchestrate shoot regeneration. Its interactions with auxin signaling and chromatin-level regulation further refine cell fate decisions, making cytokinin a central determinant of shoot meristem establishment during in vitro regeneration.

3.5 Hormone Crosstalk and Gene Regulatory Modules

Auxin and cytokinin signaling pathways function as central regulators of plant development, yet their activities are not independent. Instead, they are intricately interconnected through multilayered regulatory interactions that operate at biosynthetic, transport, signaling, and transcriptional levels. During in vitro regeneration, this hormonal crosstalk is particularly crucial because the balance between auxin and cytokinin determines whether cells adopt root, shoot, or undifferentiated callus fates. Rather than acting as simple antagonists, auxin and cytokinin establish dynamic feedback networks that create hormone-responsive gene modules capable of driving robust, switch-like developmental transitions (Su et al. 2009). One of the primary points of interaction between auxin and cytokinin occurs at the level of hormone biosynthesis. Auxin has been shown to negatively regulate cytokinin production by repressing the expression of *ISOPENTENYL TRANSFERASE* (*IPT*) genes, which encode key enzymes catalyzing the rate-limiting step in cytokinin biosynthesis. Studies in *Arabidopsis* have demonstrated that auxin-induced *ARFs*, particularly *ARF5*/MP, can indirectly suppress *IPT* gene expression, thereby reducing cytokinin accumulation in specific tissues (Olatunji et al. 2017). This repression reinforces auxin-dominant developmental programs such as root formation. Conversely, cytokinin can modulate auxin biosynthesis by influencing *YUC* gene expression, demonstrating that biosynthetic control operates bidirectionally (Raspor et al. 2021). Through these reciprocal effects on hormone production, cells can locally amplify one signaling pathway while attenuating the other. Another major regulatory node in auxin-cytokinin interaction involves auxin transport. Directional auxin flow, mediated primarily by *PIN* efflux carriers, establishes concentration gradients that function as positional signals during organogenesis. Cytokinin has been shown to regulate both the transcription and subcellular localization of *PIN* proteins. For example, cytokinin signaling can promote the internalization and degradation of *PIN1*, thereby altering auxin distribution patterns within regenerating tissues (Bielach et al. 2017). By modulating auxin transport capacity, cytokinin effectively reshapes auxin maxima and redirects developmental trajectories. This regulation ensures that cytokinin-induced shoot meristem formation occurs in spatial domains where auxin concentration is appropriately configured. At the signaling level, auxin and cytokinin pathways converge through cross-regulation of downstream transcription factors. Auxin signaling operates via *TIR1/AFB* receptors and *ARF* transcription factors, while cytokinin signaling is mediated by *AHK* receptors, *AHP* phosphotransfer proteins, and type-B *ARRs*. Type-B *ARRs* can directly regulate genes involved in auxin transport and signaling, including members of the *PIN* and *AUX/IAA* families, thereby integrating cytokinin signals into auxin-responsive transcriptional programs (Argyros et al. 2008). Similarly, *ARFs* can influence cytokinin signaling components by modulating the expression of type-A *ARR* genes, which serve as negative feedback regulators of cytokinin responses. These interactions create tightly interlocked feedback loops that coordinate hormonal outputs.

The concept of hormone-responsive gene modules has emerged from system-level analyses of regeneration. Transcriptomic profiling of callus tissues undergoing organogenesis has revealed coordinated expression of auxin- and cytokinin-regulated gene clusters. These modules often include *ARFs*, *ARRs*, *LBD* transcription factors, *WUS*, *STM*, and *PLT* genes. Rather than functioning in linear cascades, these components participate in interconnected regulatory circuits that respond collectively to hormonal ratios (Su et al. 2011). Such modular organization provides robustness, ensuring that developmental outcomes are maintained despite fluctuations in hormone levels. Mathematical modeling has further illuminated the emergent properties of auxin-cytokinin crosstalk. Computational simulations incorporating feedback between *ARFs* and *ARRs* have demonstrated that the network architecture can generate bistable states—stable configurations corresponding to root-promoting or shoot-promoting developmental programs (De Rybel et al.

2014). In a bistable system, small changes in hormone ratios can trigger a decisive and irreversible switch in gene expression patterns, enabling cells to transition sharply between alternative fates. For example, high auxin-to-cytokinin ratios stabilize *ARF*-driven transcriptional networks that activate *PLT* genes and root identity programs. In contrast, elevated cytokinin signaling activates *ARR*-mediated induction of *WUS* and *STM*, establishing shoot meristem identity. Feedback loops are central to stabilizing these bistable states. In root-promoting conditions, auxin-induced *ARFs* activate *LBD* transcription factors that enhance local auxin biosynthesis and transport, reinforcing auxin maxima. At the same time, auxin suppresses cytokinin biosynthesis and signaling components, preventing interference from shoot-promoting pathways. Conversely, during shoot regeneration, cytokinin-induced type-B *ARRs* activate *WUS* expression, which participates in a feedback loop with *CLAVATA* signaling to maintain stem cell populations.

Cytokinin also induces type-A *ARRs*, which attenuate excessive cytokinin responses and refine spatial boundaries of meristem formation (Kieber and Schaller 2018). These multilayered feedback interactions prevent ambiguous intermediate states and promote stable organ identity. Epigenetic regulation adds another dimension to auxin-cytokinin crosstalk. Chromatin remodeling complexes influence the accessibility of hormone-responsive genes, thereby modulating sensitivity to hormonal cues. During regeneration, removal of repressive histone marks such as *H3K27me3* from *WUS* or *LBD* loci enhances their inducibility by cytokinin or auxin, respectively. This epigenetic priming allows hormonal signals to elicit rapid and coordinated transcriptional changes (Kim 2019). Consequently, hormone crosstalk is embedded within a broader regulatory landscape that integrates transcriptional and chromatin-level controls. Importantly, auxin-cytokinin interactions are not strictly antagonistic; in certain contexts, they act synergistically. For instance, initial callus formation often requires both auxin-induced cell proliferation and cytokinin-mediated competence acquisition. Sequential exposure to auxin-rich and cytokinin-rich media mimics the natural progression from dedifferentiation to shoot initiation. Thus, the outcome of hormone interaction depends on concentration, timing, spatial distribution, and developmental context.

3.6 Regeneration Mechanisms in Model Plants

Model plants are widely used in regeneration studies because they provide simple and well-characterized systems for understanding complex biological processes. They generally have short life cycles, possess small genomes, and can be easily grown under laboratory conditions, which allow experiments to be performed quickly and efficiently. In addition, they have well-developed genetic resources and established tissue culture protocols. So, the regeneration studies in *Arabidopsis*, *Oryza sativa*, and *Zea mays* are as follows.

3.6.1 *Regeneration Mechanisms in* Arabidopsis

Arabidopsis thaliana has served as the primary model system for elucidating the molecular and hormonal mechanisms underlying plant regeneration. Its small genome, extensive mutant collections, and well-characterized developmental pathways have enabled detailed dissection of auxin-cytokinin interactions during in vitro organogenesis and somatic embryogenesis. Most of our current understanding of regeneration-associated gene regulatory networks originates from studies in *Arabidopsis*. In *Arabidopsis*, regeneration typically follows a two-step protocol involving callus induction on auxin-rich medium, followed by shoot induction on cytokinin-rich medium. Importantly, transcriptomic analyses have demonstrated that callus formed from various explants does not represent a completely dedifferentiated mass, but instead resembles lateral root primordium tissue at the molecular level (Sugimoto et al. 2010). This finding shifted the classical view of dedifferentiation and revealed that regeneration reactivates existing developmental programs rather than creating entirely new ones. Auxin signaling plays a central role in callus formation. *ARF7* and *ARF19* activate *LBD* transcription factors such as *LBD16* and *LBD29*, which promote cell proliferation and reprogramming (Fan et al. 2012). During the transition to shoot regeneration, cytokinin signaling becomes dominant. Type-B *ARRs* (*ARR1*, *ARR10*, *ARR12*) induce *WUS*, a master regulator *SAM* identity (Gordon et al. 2009). *WUS* establishes a feedback loop with *CLAVATA* genes to maintain stem cell homeostasis. Epigenetic regulation is also crucial in *Arabidopsis* regeneration. Chromatin remodeling complexes and histone modifications regulate the accessibility of key developmental genes such as *WUS* and *LBDs*. For instance, removal of *H3K27me3* repressive marks facilitates transcriptional activation during shoot induction (Kim 2019). Mutations in chromatin modifiers often result in reduced regeneration efficiency, demonstrating that epigenetic plasticity underlies developmental competence. Thus, *Arabidopsis* provides a mechanistic framework in which regeneration is governed by hormone-controlled transcriptional modules, spatial auxin gradients, cytokinin-mediated meristem induction, and chromatin remodeling. This model has become foundational for translating regeneration knowledge to crop species.

3.6.2 *Regeneration Mechanisms in Rice*

Rice (*Oryza sativa*) is one of the most important staple crops worldwide and serves as a model monocot species for functional genomics and biotechnology. Although rice regeneration shares core hormonal principles with *Arabidopsis*, important monocot-specific regulatory features influence regeneration efficiency. In rice, regeneration typically begins with auxin-induced callus formation from mature embryos or scutellar tissues using high concentrations of 2,4-D. Similar to *Arabidopsis*, auxin promotes cell proliferation and acquisition of embryogenic competence. However, rice callus formation is highly dependent on genotype and tissue source, with japonica varieties generally exhibiting higher regeneration capacity than indica varieties. Transition from callus to shoot regeneration requires cytokinin-rich media, which induces shoot meristem formation. Rice homologs of *WUS* and *STM* are reported to be upregulated during shoot induction, indicating conservation of meristem regulatory pathways (Zhang et al. 2021). However, rice also exhibits monocot-specific regulatory elements that influence gene expression patterns during regeneration. Recent transcriptomic and genome-wide studies have revealed conserved gene modules involving auxin biosynthesis (*YUC* genes), cytokinin signaling components (*OsRRs*), and stem cell regulators. At the same time, epigenetic constraints appear more pronounced in cereals. DNA methylation patterns and chromatin states strongly influence embryogenic potential, while differential epigenetic landscapes between genotypes partly explain variability in regeneration efficiency. Furthermore, rice somatic embryogenesis often requires stress-related signaling inputs, including reactive oxygen species and wound responses, suggesting integration of hormonal and stress pathways. These regulatory layers make rice regeneration more complex than the *Arabidopsis* model but provide opportunities for targeted genetic improvement. Overall, rice regeneration reflects both conservation of core auxin-cytokinin regulatory mechanisms and divergence due to monocot-specific genomic and epigenetic contexts.

3.6.3 Regeneration Mechanisms in Maize

Maize (*Zea mays*) represents unique challenges and opportunities in plant regeneration research. Unlike *Arabidopsis*, maize regeneration is highly genotype-dependent, with certain inbred lines displaying strong embryogenic potential, while others are recalcitrant. This variability has made maize an important system for identifying genetic determinants of regeneration competence. Somatic embryogenesis is the primary route of regeneration in maize. Immature embryos are commonly used as explants, and high auxin concentrations (typically 2,4-D) induce the formation of embryogenic callus. A key breakthrough in maize regeneration came with the identification of morphogenic regulators such as *BBM* and *WUS2*. *BBM*, an *AP2/ERF* transcription factor, promotes embryogenic cell fate and enhances somatic embryo formation. Overexpression of *BBM* significantly improves transformation and regeneration efficiency in maize and other cereals (Lowe et al. 2016). Similarly, *WUS2* expression stimulates meristem formation and synergizes with *BBM* to enhance regeneration capacity. The combined overexpression of *BBM* and *WUS2* has revolutionized maize transformation systems by enabling rapid and genotype-independent regeneration. These findings highlighted that manipulation of key developmental regulators can overcome inherent regeneration barriers. At the molecular level, maize regeneration involves activation of auxin-responsive genes, chromatin remodeling factors, and stress-associated pathways. However, compared to *Arabidopsis*, the regulatory network appears more dependent on specific morphogenic regulators rather than solely on hormone ratios (Andorf et al. 2019). Epigenetic reprogramming also plays a major role, as embryogenic competence correlates with open chromatin states and reduced DNA methylation at developmental loci. Genome-wide expression analyses have revealed both conserved modules (*ARFs*, *PINs*, cytokinin response regulators) and maize-specific regulatory components. Importantly, insights from maize have influenced regeneration strategies in other cereals, including wheat and sorghum (Ikeuchi et al. 2016). Thus, maize regeneration illustrates how core hormonal frameworks are conserved, yet successful crop biotechnology often requires direct manipulation of master transcription factors such as *BBM* and *WUS2* to bypass genotype-dependent constraints. The key differences between these model organisms are in Table 3.1.

3.7 Conclusion

Plant regeneration is one of the most remarkable evidence of developmental changes in biological systems. The development of completely organized tissues and functional meristems can result from processes initiated by factors such as wounding or exposure to plant hormones. Over time, understanding of this phenomenon has improved significantly. The conventional auxin-cytokinin ratio model provided an important early foundation, but regeneration is now recognized as a far more complex process involving linked hormonal signaling channels, tightly controlled gene networks, and dynamic epigenetic changes. Regeneration proceeds through a sequence of stages, including dedifferentiation, acquisition of competence, and redifferentiation. During the initial phase, auxin signaling through *TIR1/AFB* receptors activates *ARF-LBD* transcriptional modules. This activation promotes cell cycle re-entry and allows differentiated cells to revert to a pluripotent state. Subsequently, cytokinin signaling plays a vital role in guiding shoot meristem development. Cytokinin activates key developmental regulators such as *WUS* and *STM*, which establish and maintain meristem identity through histidine kinase receptors and *ARRs*. Rather than functioning independently, auxin and cytokinin pathways interact extensively at biosynthetic, transport, transcriptional, and chromatin levels to ensure that developmental transitions are stable and precisely controlled. Equally important is the epigenetic landscape within which these signals operate. The ability of cells to respond to hormonal signals depends heavily on chromatin accessibility and epigenetic setup. Changes in histone marks, DNA methylation patterns, and chromatin remodeling activities determine whether regeneration-associated genes can be activated at the right time and place. In this way, regeneration reflects not only hormonal signaling but also a coordinated restructuring of the genome's functional state. Comparative studies in *Arabidopsis thaliana*, *Oryza sativa*, and *Zea mays* reveal both shared common regulatory principles and significant species-specific variations. While *Arabidopsis* provides a clear mechanistic foundation, cereal crops such as rice and maize often display stronger genotype dependence and epigenetic barriers. In these species, enhancing regeneration frequently requires

Table 3.1 Differences between the regeneration mechanisms of model plants: *Arabidopsis thaliana*, *Oryza sativa*, and *Zea mays*

Features	*Arabidopsis thaliana*	*Oryza sativa*	*Zea mays*	References
Model organism	Primary dicot model plant for regeneration studies	Monocot model plant and major cereal crop	Recalcitrant monocot crop with genotype dependence	Sugimoto et al. (2010); Ikeuchi et al. (2016)
Explant source	Root, hypocotyl, leaf tissues	Mature embryo, scutellum	Immature embryos	Fehér (2019)
Callus induction hormone	2,4-D or IAA	2,4-D	2,4-D	Sugimoto et al. (2010)
Nature of callus	Resembles lateral root primordium	Embryogenic callus	Type I and type II embryogenic callus	Sugimoto et al. (2010); Fehér (2019)
Key auxin regulatory genes	*ARF5*, *ARF7*, *ARF19*; *LBD16/29*	*OsARFs*, *OsIAA* family	*ZmARFs*; auxin-responsive TFs	Fan et al. (2012); Su et al. (2011)
Shoot induction hormone	High cytokinin	High cytokinin	Cytokinin + morphogenic regulators	Kieber and Schaller (2018)
Key cytokinin regulators	*ARR1*, *ARR10*, *ARR12 (type-B ARRs)*	*OsRR* (*ARR* homologs)	*ZmRR*	Hwang et al. (2012); Kieber and Schaller (2018)
Meristem regulators	*WUS*, *STM*	*OsWUS*, *OSH1* (*STM* homolog)	*WUS2*, *WOX* family	Gordon et al. (2009)
DR genes	*WUS*, *PLT*, *LBD*	*BBM*-like, *SERK*, *WOX*	*BBM*, *WUS2*	Lowe et al. (2016)
Epigenetic regulation	*H3K27me3* removal at *WUS*; chromatin remodeling critical	DNA methylation and histone modifications affect competence	Chromatin accessibility strongly linked to embryogenic competence	Xu and Huang (2014)
Genotype dependency	Low (most ecotypes regenerate)	Moderate	High (strong genotype effect)	Fehér (2019)
Regeneration pathway	Mostly organogenesis	Somatic embryogenesis + organogenesis	Primarily somatic embryogenesis	Ikeuchi et al. (2016)
Hormone crosstalk	Well-characterized auxin-cytokinin feedback loops	Conserved but monocot-specific modifications	Hormone networks plus strong DR overexpression requirement	Su et al. (2011)
Transcriptomic insights	Regeneration mimics lateral root development	Activation of cell cycle + embryogenesis pathways	Activation of morphogenic regulators central to competence	Sugimoto et al. (2010)
Transformation efficiency	High	Moderate	Historically low; improved with *BBM/WUS2*	Lowe et al. (2016)

targeted activation of morphogenic regulators like *BBM* and *WUS2*. Plant regeneration should be viewed as a systems-level developmental program that integrates hormonal gradients, gene regulatory networks, feedback mechanisms, and chromatin dynamics. Expanding mechanistic understanding of these interrelated modules will be crucial for overcoming regeneration recalcitrance and for improving transformation and genome editing strategies in major crop species.

References

Andorf C, Beavis WD, Hufford M, Smith S, Suza WP, Wang K, Woodhouse M, Yu J, Lübberstedt T (2019) Technological advances in maize breeding: past, present and future. Theor Appl Genet 132(3):817–849

Argyros RD, Mathews DE, Chiang YH, Palmer CM, Thibault DM, Etheridge N, Argyros DA, Mason MG, Kieber JJ, Schaller GE (2008) Type B response regulators of Arabidopsis play key roles in cytokinin signaling and plant development. Plant Cell 20(8):2102–2116

Banda J, Bellande K, von Wangenheim D, Goh T, Guyomarc'h S, Laplaze L, Bennett MJ (2019) Lateral root formation in Arabidopsis: a well-ordered LRexit. Trends Plant Sci 24(9):826–839

Bidabadi SS, Jain SM (2020) Cellular, molecular, and physiological aspects of in vitro plant regeneration. Plants 9(6):702

Bielach A, Hrtyan M, Tognetti VB (2017) Plants under stress: involvement of auxin and cytokinin. Int J Mol Sci 18(7):1427

Che P, Lall S, Nettleton D, Howell SH (2006) Gene expression programs during shoot, root, and callus development in Arabidopsis tissue culture. Plant Physiol 141(2):620–637

De Rybel B, Adibi M, Breda AS, Wendrich JR, Smit ME, Novák O, Yamaguchi N, Yoshida S, Van Isterdael G, Palovaara J, Nijsse B, Weijers D (2014) Integration of growth and patterning during vascular tissue formation in Arabidopsis. Science 345(6197):1255215

Fan M, Xu C, Xu K, Hu Y (2012) LATERAL ORGAN BOUNDARIES DOMAIN transcription factors direct callus formation in Arabidopsis regeneration. Cell Res 22(7):1169–1180

Fehér A (2019) Callus, dedifferentiation, totipotency, somatic embryogenesis: what these terms mean in the era of molecular plant biology? Front Plant Sci 10:536

Gordon SP, Chickarmane VS, Ohno C, Meyerowitz EM (2009) Multiple feedback loops through cytokinin signaling control stem cell number within the Arabidopsis shoot meristem. Proc Natl Acad Sci 106(38):16529–16534

Hwang I, Sheen J, Müller B (2012) Cytokinin signaling networks. Annu Rev Plant Biol 63:353–380

Ikeuchi M, Ogawa Y, Iwase A, Sugimoto K (2016) Plant regeneration: cellular origins and molecular mechanisms. Development 143(9):1442–1451

Kepinski S, Leyser O (2005) The Arabidopsis F-box protein TIR1 is an auxin receptor. Nature 435(7041):446–451

Kieber JJ, Schaller GE (2018) Cytokinin signaling in plant development. Development 145(4):dev149344

Kim JH (2019) Chromatin remodeling and epigenetic regulation in plant DNA damage repair. Int J Mol Sci 20(17):4093

Kutschera U, Niklas KJ (2013) Cell division and turgor-driven stem elongation in juvenile plants: a synthesis. Plant Sci 207:45–56

Leyser O (2018) Auxin signaling. Plant Physiol 176(1):465–479

Li J, Zhang Q, Wang Z, Liu Q (2024) The roles of epigenetic regulators in plant regeneration: exploring patterns amidst complex conditions. Plant Physiol 194(4):2022–2038

Lowe K, Wu E, Wang N, Hoerster G, Hastings C, Cho MJ, Scelonge C, Lenderts B, Chamberlin M, Cushatt J, Wang L, Gordon-Kamm W (2016) Morphogenic regulators Baby boom and Wuschel improve monocot transformation. Plant Cell 28(9):1998–2015

Olatunji D, Geelen D, Verstraeten I (2017) Control of endogenous auxin levels in plant root development. Int J Mol Sci 18(12):2587

Petrášek J, Friml J (2009) Auxin transport routes in plant development. Dev 136(16):2675–2688

Raspor M, Motyka V, Kaleri AR, Ninković S, Tubić L, Cingel A, Ćosić T (2021) Integrating the roles for cytokinin and auxin in de novo shoot organogenesis: from hormone uptake to signaling outputs. Int J Mol Sci 22(16):8554

Skoog F (1957) Chemical regulation of growth and organ formation in plant tissue cultured in vitro. Symp Soc Exp Biol 11:118–131

Su YH, Zhao XY, Liu YB, Zhang CL, O'Neill SD, Zhang XS (2009) Auxin-induced WUS expression is essential for embryonic stem cell renewal during somatic embryogenesis in Arabidopsis. Plant J. 59(3):448–460

Su YH, Liu YB, Zhang XS (2011) Auxin-cytokinin interaction regulates meristem development. Mol Plant 4(4):616–625

Sugawara S, Hishiyama S, Jikumaru Y, Hanada A, Nishimura T, Koshiba T, Zhao Y, Kamiya Y, Kasahara H (2009) Biochemical analyses of indole-3-acetaldoxime-dependent auxin biosynthesis in Arabidopsis. Proc Natl Acad Sci 106(13):5430–5435

Sugimoto K, Jiao Y, Meyerowitz EM (2010) Arabidopsis regeneration from multiple tissues occurs via a root development pathway. Dev Cell 18(3):463–471

Sugimoto K, Temman H, Kadokura S, Matsunaga S (2019) To regenerate or not to regenerate: factors that drive plant regeneration. Curr Opin Plant Biol 47:138–150

To JP, Deruère J, Maxwell BB, Morris VF, Hutchison CE, Ferreira FJ, Schaller GE, Kieber JJ (2007) Cytokinin regulates type-a Arabidopsis response regulator activity and protein stability via two-component phosphorelay. Plant Cell 19(12):3901–3914

Vissenberg K, Claeijs N, Balcerowicz D, Schoenaers S (2020) Hormonal regulation of root hair growth and responses to the environment in Arabidopsis. J Exp Bot 71(8):2412–2427

Xu L, Huang H (2014) Genetic and epigenetic controls of plant regeneration. Curr Top Dev Biol 108:1–33

Xu C, Chang P, Guo S, Yang X, Liu X, Sui B et al (2024) Transcriptional activation by WRKY23 and derepression by removal of bHLH041 coordinately establish callus pluripotency in Arabidopsis regeneration. Plant Cell 36(1):158–173

Zhang TQ, Chen Y, Liu Y, Lin WH, Wang JW (2021) Single-cell transcriptome atlas and chromatin accessibility landscape reveal differentiation trajectories in the rice root. Nat Commun 12(1):2053

Chapter 4
Epigenetic Gatekeepers of the Totipotent State

Abstract Somatic embryogenesis represents a salient demonstration of plant cellular plasticity, yet the mechanisms enabling differentiated somatic cells to reacquire totipotency remain incompletely understood. Increasing evidence identifies epigenetic regulation as a central determinant of regenerative competence, governing whether developmental regulatory genes remain transcriptionally inactive or become accessible for activation. Epigenetic regulation includes DNA methylation, histone modifications, and ATP-dependent chromatin remodeling coordinate transcriptional reprogramming during in vitro regeneration. Additionally, epigenetic regulation has an important role in modulating plant cell totipotency, emphasizing how dynamic methylation changes and histone marks influence callus formation, somatic embryogenesis, and plantlet regeneration. The chapter further explores epigenetic control of developmental regulatory gene accessibility, highlighting the functions of Polycomb Repressive Complex 2 (*PRC2*), SWItch/Sucrose Non-Fermentable (*SWI/SNF*) chromatin remodeling complexes, and histone methyltransferases in regulating key morphogenic regulators such as *LEAFY COTYLEDON* (*LEC*), *WUSCHEL* (*WUS*), and *BABY BOOM* (*BBM*). Moreover, during in vitro regeneration, epigenetic reprogramming addresses how prolonged tissue culture induces both reversible epigenomic reprogramming and somaclonal variations. Finally, strategies for overcoming epigenetic silencing in recalcitrant species are evaluated, including chemical modulators, manipulation of chromatin regulators, and optimized culture systems. By integrating recent insights into chromatin dynamics, this chapter underscores that understanding the plant epigenome is important for improving regeneration rates and advancing biotechnology in species that are traditionally difficult to transform.

Keywords Epigenetic regulation · Histone modification · PRC2 · Recalcitrant species · SWI/SNF complexes · Totipotency

P. Kumar, *Totipotency in Plants*, SpringerBriefs in Plant Science,
https://doi.org/10.1007/978-981-92-1631-4_4

4.1 Introduction

Somatic embryogenesis, an outstanding manifestation of cellular plasticity and regenerative potential, has attracted sustained scientific attention for more than seven decades. Despite substantial advances, a comprehensive understanding of the mechanisms by which differentiated somatic cells reacquire an embryogenic program remains elusive. The establishment of embryogenic competence is controlled by an intricate network of regulatory pathways involving hormonal cues, stress-responsive signaling, cell wall remodeling, extensive transcriptional reprogramming, and multilayered epigenetic regulation. Among these, epigenetic regulation has emerged as a central determinant of whether developmental regulatory genes become accessible for activation or remain transcriptionally inactive (Carra et al. 2024). Plant tissue culture enables clonal propagation and production of true-to-type plants. However, cultured cells frequently undergo genetic and epigenetic alterations during prolonged in vitro subculturing, ultimately causing variability among regenerated plantlets through both genetic and epigenetic routes. In vitro genetic variation events are generally considered relatively infrequent and can result from mutations, chromosome doubling, and structural aberrations. In contrast, epigenetic modifications can reversibly activate or repress functional genes and thereby produce heritable phenotypic changes without altering the nucleotide sequence (Azizi et al. 2020). Major contributors to this variability include DNA methylation and demethylation, post-translational histone modifications, chromatin compaction, and ATP-dependent chromatin remodeling. The plant epigenome, defined as the collection of heritable chromatin marks superimposed on the DNA sequence, governs gene expression and phenotype without altering nucleotide composition. These marks consist primarily of chemical modifications to DNA bases and histone protein tails. Through their influence on chromatin structure, these modifications determine whether genomic regions adopt euchromatin or restrictive states for transcription (Wójcikowska et al. 2020). In addition, epigenetic modifications lead to somaclonal variation by altering gene expression programs and developmental competence in regenerating tissues, often in association with cytological abnormalities such as aneuploidy, reflecting disturbances arising during mitotic division. Recurrent epigenomic reprogramming during culture can therefore increase phenotypic divergence among regenerated lines, highlighting the dual role of epigenetic mechanisms as both facilitators and barriers to totipotency (Majumder et al. 2025).

Accumulating evidence indicates that successful acquisition of the totipotent state requires coordinated chromatin remodeling events that reactivate embryogenic and meristematic pathways. Polycomb Repressive Complex 2 (*PRC2*), SWItch/Sucrose Non-Fermentable (*SWI/SNF*) chromatin remodeling assemblies and multiple classes of histone methyltransferases play decisive roles in shaping the accessibility of developmental regulatory genes such as *LEAFY COTYLEDON* (*LEC*), *WUSCHEL* (*WUS*), *BABY BOOM* (*BBM*), and related regulators. Depending on developmental pathways, these factors can stabilize repression to maintain cell identity and, conversely, permit transcriptional reprogramming when epigenetic barriers are relieved (Huang et al. 2025).

Understanding how such epigenetic gatekeepers activate is particularly important for recalcitrant species, in which regeneration competence is limited. Strategies aimed at overcoming epigenetic silencing through chemical modulators, targeted manipulation of chromatin regulators, or culture regimes that favor transcriptionally active euchromatin states are increasingly observed to enhance transformation and regeneration efficiency. This chapter examines how epigenetic pathways modulate plant cell totipotency, with special emphasis on the mechanisms by which epigenetic regulators control developmental regulatory gene accessibility and how these constraints can be mitigated in recalcitrant species that are difficult to regenerate.

4.2 Role of Epigenetic Regulation in Modulating Plant Cell Totipotency

Epigenetic regulation constitutes a central regulatory framework leading to plant cell totipotency, particularly during callus induction, somatic embryogenesis, and whole-plant regeneration in plant tissue culture. Dynamic modifications such as DNA methylation, histone post-translational modifications, and ATP-dependent chromatin remodeling coordinate transcriptional reprogramming by activating or repressing developmental regulators (Sivanesan et al. 2022). These complex mechanisms function as molecular gatekeepers, determining whether differentiated somatic cells retain their identity or reacquire embryogenic competence.

4.2.1 Histone Marks and DNA Methylation Dynamics

4.2.1.1 Regulation of Gene Expression and Regeneration

Among various epigenetic mechanisms, DNA methylation has been widely recognized as a critical factor influencing phenotypic variation and regulating regenerative pathways in plants as shown in Table 4.1. DNA methylation, defined as the addition of a methyl group to cytosine residues in CG, CHG, and CHH nucleotide configuration, represents a major epigenetic mark influencing transcriptional activity in plants as shown in Fig. 4.1. Histone tails, in contrast, can undergo diverse post-translational modifications, including acetylation, methylation, phosphorylation, and ubiquitination, collectively referred to as histone marks. Alterations in global and locus-specific methylation patterns significantly affect regeneration efficiency and morphological stability during plant tissue culture (Shimooka et al. 2013). Prolonged in vitro subculture has been associated with shifts in global DNA methylation levels, often correlating with reduced regenerative capacity and somaclonal variation. The methylation landscape also varies depending on the explant

Table 4.1 Overview of epigenetic gatekeepers controlling somatic embryogenesis and organogenesis in plants

Plant species		Epigenetic gatekeepers	Targeted developmental genes	In vitro observations	References
Modal plant	*Arabidopsis thaliana*	*H3K4me3*, *H3K9ac*	*WUS*	Increases shoot regeneration	Li et al. (2011)
		H3K9me2	*WUS*	Decreases shoot regeneration	
		H3K9me2, *H3K4me3*, *H3K9ac*	*CUC1*, *CUC2*	Increases callus and shoot induction	Song et al. (2017)
		H3K4me2	*LDL3*	Decreases shoot induction	Ishihara et al. (2019)
		H3K27 methyltransferase	–	Promote somatic embryogenesis through cell dedifferentiation	Chanvivattana et al. (2004)
		H3K27me3	–	Induction of callus-like structures	Bouyer et al. (2011)
		H3K27me3	*PIN6*	Affects root development	Cazzonelli et al. (2013)
		H3K27me3	Silencing of leaf regulatory genes and activation of auxin pathway respectively	Helps in callus formation from leaf through cell dedifferentiation	He et al. (2012)
		Loss of DNA methylation, met1 mutant	*WUS*, *ARF3*	Increases shoot regeneration	Li et al. (2011)
		Loss of *KYP*, *JMJ14*, *MET1*, and *HAC1*	*WUS*	Changes shown in shoot regeneration	
		drm1 drm2 mutant, *drm1 drm2 cmt3* mutant	*LEC1*, *LEC2*, *BBM*	Regulate somatic embryogenesis process	Grzybkowska et al. (2018)
		hda6 hda19 mutant	*LEC1*, *FUS3*, *ABI3*	Formation of embryo-like structures	Tanaka et al. (2008)

Woody and perennial crops	*Coffea canephora*	*H3K27me3*	*LEC1, BBM1*	Suppressed somatic embryogenesis	Nic-Can et al. (2013)
		H3K4me3, H3K36me2	*BBM1*	Enhances somatic embryogenesis	
		DNA demethylation	*LEC1, BBM1*	Enhanced somatic embryo formation	
	Gossypium hirsutum	Promoter hypomethylation	*WOX* genes	Increased SE production	Li et al. (2019)
Monocot species	*Oryza sativa* (rice)	Increased *H3K4me3* & reduced *H3K27me3*	*BBM, ABI3*	Callus-specific activation	Zhang et al. (2017)
		Loss of *H3K9* methylation	*Tos17*	Reduced callus proliferation	Ding et al. (2007)
Gymnosperms	*Pinus radiate*	*H4ac* and *H3K4me3*	–	Maintenance of somatic embryogenesis and adventitious shoot potential	Valledor et al. (2010)
	Picea abies, P. Sylvestris	Histone deacetylase	Decrease in *LEC1*-type HAP3A	Affects embryo development	Uddenberg et al. (2011)

(continued)

Table 4.1 (continued)

Plant species		Epigenetic gatekeepers	Targeted developmental genes	In vitro observations	References
Herbaceous crops and horticultural species	*Nicotiana tabacum*	*H3K9ac* and *H3K14ac*	Triggers *pRb/E2F* targeted *RNR2* and *PCNA*	Activation of cell cycle genes during protoplast culture	Williams et al. (2003)
	Agave spp.	Increases *H3K4me3* and *H3K9me2*	Increases *AtqKNOX1* and decreases *AtqKNOX1* respectively	Activation of *KNOX* genes in vitro	De-la-Peña et al. (2012)
	Boesenbergia rotunda	Decrease in CG and CHG methylation	*WUS*, *SERK*, *BBM*, *LEC*	Increases in somatic embryogenesis production	Karim et al. (2018)
	Lotus japonicus	DNA methylation	Reactivation *LORE1a*	Alteration in micropropagation	Fukai et al. (2010)
	Fragaria × *ananassa*	Decline in *FaMET1*	–	Epigenetic variation during micropropagation	Chang et al. (2009)

ABI3 Abscisic Acid Insensitive 3, ARF3 AUXIN Response Factor 3, BBM BABY BOOM, cmt3 CHROMOMETHYLASE 3, CUC1 CUP-SHAPED COTYLEDON 1, drm1/drm2 Domains Rearranged Methyltransferase 1/2, FUS3 FUSCA3, HAC1 Histone Acetyltransferase of the CBP FAMILY 1, HAP3A HEME Activator Protein 3A, H4ac Histone H4 Acetylation, hda6/hda9 Histone Deacetylase 6/HISTONE DEACETYLASE 9, H3K9ac Histone H3 lysine 9 acetylation, H3K14ac Histone H3 Lysine 14 Acetylation, H3K27me3 Histone H3 Lysine 27 Trimethylation, H3K36me2 Histone H3 Lysine 36 Dimethylation, H3K4me3, Histone H3 Lysine 4 Trimethylation, H3K9me2 Histone H3 Lysine 9 Dimethylation, JMJ14, JUMONJI Domain-Containing Protein 14, KNOX1 KNOTTED1-like homeobox gene (AtqKNOX1), KYP kryptonite, LDL3 LYSINE-Specific Demethylase-Like 3, LEC Leafy Cotyledon (LEC1, LEC2), LORE1a, Lotus Retrotransposon 1a, MET1 Methyltransferase 1, PCNA PIN6 PIN-FORMED 6, pRb/E2F Retinoblastoma protein/E2F transcription factor complex, RNR2 Ribonucleotide Reductase 2, Tos17 Tobacco retrotransposon 17, WOX WUSCHEL-Related Homeobox, WUS WUSCHEL, SERK SOMATIC Embryogenesis Receptor-Like Kinase

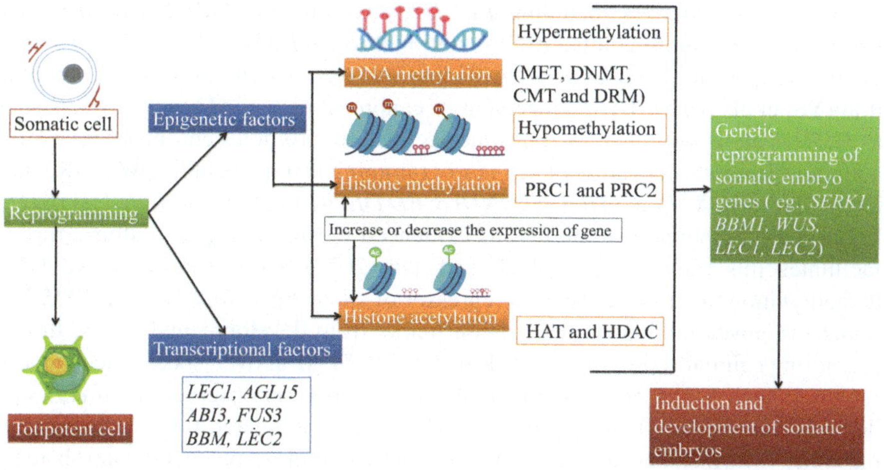

Fig. 4.1 Epigenetic control of plant cell reprogramming and totipotency (*ABI3* abscisic acid insensitive 3, *AGL15* AGAMOUS-like 15, *BBM* BABY BOOM, *CMT* chromomethylase, *DNMT* DNA methyltransferase, *DRM* domains rearranged methyltransferase, *FUS3 FUSCA 3*, *HAT* histone acetyltransferase, *HDAC* histone deacetylase, *LEC2* LEAFY COTYLEDON 2, MET, *PRC* Polycomb Repressive Complex, *SERK1* somatic embryogenesis receptor-like kinase 1, *WUS*, *WUSCHEL*)

source and developmental stage. Notably, dynamic DNA demethylation is frequently required for somatic embryogenesis induction, whereas controlled remethylation is essential for stable plantlet regeneration (Chen et al. 2020). Species-specific differences have been reported in *Daucus carota*, where embryogenic cells exhibit elevated methylation levels, while lower methylation favors embryogenic callus formation in *Eleutherococcus senticosus* (Kumar and Van Staden 2017). These findings indicated that precise methylation reprogramming underlies embryogenic differentiation during both somatic and zygotic embryogenesis. Recent studies emphasize that *LEC2* has emerged as a central regulator that coordinates chromatin remodeling and activates key totipotency-associated genes via integrated epigenetic and hormonal pathways. *LEC2* promotes chromatin accessibility by stimulating RNA-directed DNA methylation (RdDM), particularly CHH methylation through the activation of *DOMAINS REARRANGED METHYLTRANSFERASE 1* and *2* (*DRM1/2*). The methylated loci are subsequently recognized by *SUVH/SDJ* complexes, which recruit AT-hook motif nuclear-localized (*AHL*) transcription factors. This recruitment facilitates histone acetylation by histone acetyltransferase (HAT) and histone deacetylases (HDACs) and chromatin relaxation, thereby enabling the transcription of embryogenic genes as shown in Fig. 4.1. This hierarchical transcriptional epigenetic framework positions *LEC2* as a molecular architect that establishes genome-wide epigenetic landscapes conducive to cellular reprogramming (Bidabadi and Jain 2020).

The redistribution of repressive histone marks such as *H3K27me3* (trimethylation of histone H3 at lysine 27), mediated by *PRC2*, allows the activation of key embryogenic regulators including *LEC1/2*, AGAMOUS-LIKE15 (*AGL15*), and *BBM*. *PRC2* maintains cellular differentiation through *H3K27me3*-mediated silencing of embryogenic regulators, whereas loss of *PRC2* activity allows dedifferentiation (Yin et al. 2025). The replacement of canonical histone H3 with variants such as *HTR15*, along with increased histone acetylation, promotes chromatin relaxation and transcriptional activation of key reprogramming factors including WUS-Related Homeobox (*WOX11*) and *WOUND INDUCED DEDIFFERENTIATION 1* (*WIND1*). Enhanced chromatin accessibility, as evidenced during protoplast culture, further facilitates this transition (Bidabadi and Jain 2020). Concurrently, active DNA demethylation by enzymes such as repressor of silencing 1 (*ROS1*) and *DEMETER* (*DME*) removes inhibitory methylation marks from developmental genes, further promoting cellular reprogramming (Rai et al. 2025). Moreover, species-specific differences, such as the more flexible epigenetic repression observed in *Torenia* epidermal cells, highlight how modulation of chromatin states determines developmental plasticity (Morinaka et al. 2021). Together, these complex epigenetic controls regulate whether differentiated plant cells retain their identity or reacquire totipotency in response to developmental or environmental cues.

4.2.1.2 Histone Modifications

Histone post-translational modifications including acetylation, methylation, phosphorylation, and ubiquitination modulate chromatin structure and transcriptional accessibility. These modifications are essential for regulating gene expression during organogenesis and embryogenic transition (Ramakrishnan et al. 2023). Among the most extensively studied marks in plants are *H3K4me2/3* (associated with active transcription), *H3K9me2* and *H3K27me3* (repressive marks), and *H3K36me3* (linked with transcriptional elongation). Genome-wide studies in *Arabidopsis thaliana* demonstrate that *H3K27me3* regulates thousands of genes, including those involved in auxin signaling and cell fate determination (Liu et al. 2019). During callus induction, *H3K27me3*-mediated repression of AUX/IAA genes is reconfigured as part of global transcriptional reprogramming. Histone modifications also interact closely with DNA methylation through methyltransferase (MET), chromomethylase (CMT), and domains rearranged methyltransferase (Lafos et al. 2011). For example, CHH hypomethylation at promoter regions through CMT has been associated with activation of *WOX* and hormone-related genes during somatic embryogenesis (Wójcikowska et al. 2020). Reduced levels of the repressive *H3K9me2* mark in certain in vitro systems have been linked to altered expression of developmental regulators such as *KNOX* (KNOTTED1-like Homeobox) genes (Le et al. 2025). Additionally, the histone variant H2A.Z and DNA methylation exhibit antagonistic interactions, with H2A.Z protecting gene bodies from excessive methylation (Zilberman et al. 2008). These findings highlighted the coordinated interplay between histone marks and DNA methylation in regulating totipotent competence.

4.2.1.3 Influence of Culture Conditions

Exogenous factors in culture media substantially modulate DNA methylation patterns. Auxins, particularly 2,4-dichlorophenoxyacetic acid (2,4-D), can induce epigenetic alterations that influence somatic embryogenesis competence. In certain species such as carrot, high concentrations of 2,4-D increase cytosine methylation and may lead to abnormal embryogenic development, whereas in others (e.g., pumpkin), methylation levels remain unaffected (Baklouti et al. 2022). Cytokinins such as 6-benzyladenine (BA) have been reported to reduce methylation by downregulating methyltransferase genes including *DRM2* (domains rearranged methyltransferase 2) and *MET1* (methyltransferase 1) (Debnath and Ghosh 2022). Furthermore, chemical agents such as 5-azacitidine (a demethylating agent) modify global methylation status and significantly impact somatic embryogenesis induction efficiency (Grzybkowska et al. 2018). Collectively, these observations underscore the dynamic and context-dependent role of DNA methylation as a regulatory gatekeeper of totipotency. In addition to epigenetic regulation, *LEC2* integrates hormonal signaling pathways, particularly auxin. Auxin not only serves as an exogenous inducer of somatic embryogenesis but also modulates chromatin accessibility and transcriptional activation of embryogenic regulators. Through interaction with *SPEECHLESS (SPCH)*, *LEC2* enhances local auxin biosynthesis by activating genes such as *TAA1* (Tryptophan Aminotransferase of Arabidopsis 1) and *YUC4* (*YUCCA 4*), thereby redirecting differentiated somatic cells toward an embryogenic fate (Wójcik 2025). Together with enhanced chromatin accessibility at promoters of morphogenic transcription factors and their integration with auxin biosynthesis, transport, and signaling pathways, these complex epigenetic controls function as molecular gatekeepers. They coordinate transcriptional and hormonal networks to unlock cellular plasticity, thereby enabling differentiated somatic cells to reacquire totipotency and initiate embryogenesis.

4.2.2 Chromatin Remodeling

ATP-dependent chromatin remodeling complexes further contribute to the establishment and maintenance of totipotency by dynamically altering nucleosome positioning and chromatin accessibility. Wounding and stress signals inherent to plant tissue culture frequently initiate large-scale chromatin reorganization, which subsequently reshapes transcriptional networks controlling regeneration (Peng et al. 2023). A comprehensive review by Miguel and Marum highlighted the dynamic nature of chromatin remodeling during cellular dedifferentiation and redifferentiation, emphasizing its transformative impact on developmental reprogramming (Bidabadi and Jain 2020). Moreover, genome-wide analyses revealed that increased chromatin accessibility at promoters and enhancers of morphogenic transcription factors is a hallmark of somatic embryogenesis induction. Hierarchical transcriptional regulatory networks, supported by transcriptionally active chromatin states, enable rapid activation of developmental regulators such as *WUS*. Functional disruption of key epigenetic regulators including *MET1* and *KRYPTONITE* (*KYP*; *H3K9 methyltransferase*) results in altered *WUS* expression and impaired shoot regeneration, underscoring the essential role of chromatin remodeling in maintaining regenerative competence (Hung et al. 2023).

4.3 Epigenetic Control of Developmental Regulatory Gene Accessibility

Epigenetic mechanisms critically determine the accessibility of developmental regulatory genes, thereby governing plant developmental transitions and regenerative responses. The balance between stable gene silencing and inducible activation is mediated through Polycomb-group complexes, ATP-dependent chromatin remodelers, and histone-modifying enzymes. Together, these molecular systems function as epigenetic gatekeepers that maintain cellular identity under normal conditions while permitting controlled reprogramming during callus formation, somatic embryogenesis, and organogenesis as shown in Table 4.1.

4.3.1 Polycomb-Mediated Repression

PRC2 is a central regulator of transcriptional repression in plants. It catalyzes trimethylation of histone H3 at lysine 27 (*H3K27me3*), a repressive epigenetic mark associated with stable gene silencing. In *Arabidopsis thaliana*, *PRC2* components such as *CURLY LEAF* (*CLF*), *SWINGER* (*SWN*), and *EMBRYONIC FLOWER2* (*EMF2*) regulate thousands of target genes, including those involved in embryogenesis, auxin signaling, and cellular differentiation (Shu et al. 2019). During callus induction and regeneration, large-scale reprogramming of *H3K27me3* occurs. Mutations in PRC2 components (i.e., *emf2*, *clf swn*) disrupt proper callus formation and induce ectopic developmental programs, highlighting *PRC2*'s role in preventing inappropriate dedifferentiation (He et al. 2012; Zheng et al. 2022). *PRC2* also regulates the embryo-to-seedling transition by maintaining repression of embryonic regulators such as *LEC1*, *FUS3*, and *ABI3* after germination (Gazzarrini and Song 2024). Thus, deposition of *H3K27me3* ensures developmental stability and prevents premature activation of embryogenic pathways. Additionally, genome-wide *H3K27me3* deposition represses key regulatory genes during vegetative growth. Loss of *H3K27* methyltransferase activity has been shown to promote cell dedifferentiation and somatic embryogenesis, indicating that *PRC2*-mediated repression functions as a molecular barrier against spontaneous reprogramming (Shen et al. 2021). Studies demonstrate that acquisition of pluripotency in cultured *Arabidopsis* tissues involves dynamic removal or redistribution of *H3K27me3* marks at promoters of morphogenic regulators such as *WUS* and *WOX* genes (Fambrini et al. 2022). This reconfiguration enables transcriptional activation necessary for regeneration.

4.3.2 *SWI/SNF* Chromatin Remodeling Complexes

SWI/SNF ATP-dependent chromatin remodeling complexes alter nucleosome positioning to regulate chromatin accessibility. In plants, these assemblies modulate transcriptional activation of developmental regulatory genes by facilitating access of transcription factors to promoter regions. Chromatin remodeling is recognized as a major mechanism underlying epigenetic variation during tissue culture and organogenic transitions (Runge et al. 2016). *SWI/SNF* complexes integrate environmental and hormonal cues with transcriptional networks. Auxin-induced callus formation requires chromatin remodeling to activate downstream regulators, and histone modifications cooperate with hormone-responsive transcription factors to establish regeneration competence (Rudolf et al. 2024). The interplay between chromatin remodeling and auxin signaling is particularly critical for activating morphogenic genes during dedifferentiation. By modulating chromatin accessibility, *SWI/SNF* remodelers create transcriptionally active chromatin states that are required for organogenesis. Reduced activity of chromatin remodelers such as PICKLE (*PKL*) has been associated with enhanced embryogenic potential, suggesting that remodeling complexes function as modulators of developmental thresholds (Fu et al. 2016). Thus, *SWI/SNF* assemblies promote the acquisition of developmental responsiveness while ensuring strict spatiotemporal control of gene expression programs.

4.3.3 *Histone Methyltransferases and Demethylases*

Histone methyltransferases and demethylases act as writers and erasers of lysine methylation marks that regulate developmental regulatory gene accessibility (Torres and Fujimori 2015). *PRC2*-mediated *H3K27me3* deposition represses a broad range of developmental genes in *Arabidopsis*, influencing over nine thousand loci involved in differentiation and hormonal signaling (Shen et al. 2021). Conversely, histone demethylases such as *JMJ14* (jumonji domain-containing protein 14; an H3K4 demethylase) modulate gene expression during regeneration; functional loss of *JMJ14* alters *WUS* expression in regenerated shoots (Jeevaraj et al. 2025). Specific histone marks including *H3K4me2/3* (active), *H3K9me2* (repressive), *H3K27me3* (repressive), and *H3K36me3* (transcription-associated) fine-tune transcriptional thresholds of developmental regulatory genes. For example, *H3K9me2* regulates *WOX4* expression in *Coffea canephora*, while reduced *H3K4me2* levels correlate with decreased shoot induction (Dong et al. 2024). Histone deacetylases such as *HDA6* and *HDA19* suppress embryonic regulators (*LEC1*, *FUS3*, *ABI3*). Their mutation results in derepression of these genes, linking histone modification to somatic embryogenesis competence. The dynamic interplay between methylation writers and erasers establishes an epigenetic equilibrium that maintains cellular identity while allowing reprogramming under inductive conditions (Tanaka et al. 2008). Reprogramming of *H3K27me3* landscapes is essential for acquiring pluripotency in cultured tissues, whereas maintenance of certain repressive marks ensures developmental stability. This epigenetic balance ultimately governs whether developmental regulatory genes remain silenced or become accessible, thereby controlling the transition between differentiated states and regenerative competence.

4.4 Overcoming Epigenetic Silencing in Recalcitrant Species

Certain plant species and genotypes exhibit recalcitrance to in vitro regeneration due to stable epigenetic silencing of developmental regulatory genes. Robust epigenetic barriers, including dynamic DNA methylation and histone modifications, are tightly linked to developmental recalcitrance in species like *Quercus suber*, often limiting the activation of totipotency networks and suppressing regenerative pathways, leading to abnormal or asynchronous embryogenesis (Wang et al. 2025). Recent studies demonstrated that targeted manipulation of the epigenome through chemical, genetic, and technological approaches can alleviate such silencing and improve regenerative competence in recalcitrant taxa.

4.4.1 Chemical and Culture-Based Approaches

4.4.1.1 Demethylating Agents

Chemical inhibitors of DNA methylation such as azacytidine (5-AzaC) and zebularine reduce cytosine methylation, thereby relieving transcriptional repression of silenced genes (Yadav et al. 2024). Application of these compounds in plant tissue culture has been shown to promote embryogenic responses, enhance protoplast-derived callus formation, and increase organogenesis efficiency by inducing hypomethylation at key loci involved in regeneration pathways. For example, in *Quercus suber*, treatment with the demethylating agent 5-AzaC significantly enhanced somatic embryogenesis output (Wang et al. 2025). In cabbage protoplast cultures, DNA (cytosine-5)-methyltransferase (DNMT) inhibitors promoted shoot organogenesis and polyploidization, indicating their role in modulating epigenetic states during regeneration (Kiełkowska and Brąszewska 2025).

4.4.1.2 Histone Deacetylase Inhibitors

Histone deacetylase inhibitors (HDACi) such as trichostatin A (TSA) promote histone hyperacetylation and open chromatin, which facilitates transcription factor access and gene expression. TSA has been shown to enhance embryogenic tissue development and improve regeneration in multiple species, including *Arabidopsis* and barley, by enhancing expression of totipotency regulators and modulating hormone responsiveness in callus cultures (Nowak et al. 2024).

4.4.1.3 Stress and Environmental Cues

Physical and cultural stressors such as osmotic stress, wounding, temperature shifts, and controlled hormone treatments can trigger chromatin relaxation and epigenetic remodeling. Stress-induced chromatin changes often reduce repressive histone methylation and DNA methylation, thereby facilitating derepression of regeneration-associated genes. Such culture conditions are widely used in protocols to elicit enhanced competence for somatic embryogenesis and organogenesis in recalcitrant explants (e.g., stress-mediated induction of totipotent gene networks) (Kowalik et al. 2026).

4.4.2 Genetic and Molecular Strategies

Genetic alteration or modulation of chromatin remodeling factors and histone modifiers provides a promising route to overcome epigenetic barriers. For example, inhibition of *PRC2* components such as *H3K27* methyltransferases can relieve repression of embryogenic regulators and facilitate callus formation and somatic embryogenesis acquisition (Wójcikowska et al. 2020). Knockdown or reduced activity of specific histone demethylases and chromatin remodelers has been shown to prime genes for activation during regeneration, enabling the establishment of pluripotent states from differentiated tissues (Suzuki and Ochi 2020). Overexpression of key morphogenic transcription factors such as *BBM* and *LEC2* has been used to improve regeneration and transformation efficiency in recalcitrant species (e.g., *Capsicum annuum* and *Populus tomentosa*), highlighting the potential of controlled gene activation to overcome epigenetic silencing and promote totipotency (Xu et al. 2024).

4.4.3 Emerging Technologies

4.4.3.1 Epigenome Editing

Epigenome editing employs engineered DNA-binding platforms such as CRISPR/Cas9 fused to epigenetic effector domains to target specific epigenetic marks at select genomic loci without altering the underlying DNA sequence. This technology enables precise activation or repression of developmental regulators by adding or removing methylation and histone marks, offering a targeted means to overcome epigenetic barriers to totipotency (Wada et al. 2020). Experimental applications in plants such as CRISPR activation of regeneration-associated genes (e.g., *SlWRKY29* in tomato) demonstrate how locus-specific epigenetic reprogramming can increase embryogenic competence and gene accessibility (Dubois and Roudier 2021; Valencia-Lozano et al. 2024).

4.4.3.2 Single-Cell Chromatin Profiling

High-resolution profiling of chromatin accessibility at the single-cell level using approaches like single-cell ATAC-Seq provides unprecedented insight into the heterogeneity of epigenetic states within plant tissues (Lu et al. 2024). Such technologies reveal distinct chromatin landscapes associated with regenerative competence, enabling identification of cell populations controlled for reprogramming and mapping of regulatory networks that underlie totipotency acquisition (Feng et al. 2022). Future adaptation of single-cell chromatin profiling in regenerating explants will refine predictive models of totipotency and guide targeted modulation strategies.

4.4.3.3 Predictive Markers of Regeneration Competence

Emerging multi-omics and epigenomic datasets are facilitating the identification of biomarkers that predict regenerative potential across species and explant types. Integration of chromatin accessibility profiles, DNA methylation maps, and transcriptional landscapes will inform robust markers of regeneration competence, enabling customized protocols to overcome recalcitrance and enhance totipotent cell induction in agriculturally important plants (Leduque et al. 2024).

4.5 Conclusion

Epigenetic gatekeepers play a decisive role in establishing and maintaining the totipotent state during somatic embryogenesis by orchestrating dynamic chromatin and DNA methylation changes. By managing changes in DNA methylation and histone modifications, these regulatory mechanisms act as molecular switches that determine whether a somatic cell remains in its current state or restarts its development through somatic embryogenesis. A central part of this process is the control of chromatin accessibility by complexes like *PRC2* and *SWI/SNF*, which are essential for activating key genes such as *LEC*, *WUS*, and *BBM*. However, these epigenetic processes have a dual nature: while they are necessary for successful plant regeneration, they are also the primary cause of somaclonal variation, which can lead to somatic variations in cloned plants. Identifying these epigenetic barriers is especially vital for recalcitrant species that are traditionally difficult to grow under in vitro conditions. Using tools such as chemical modulators and optimized culture methods to promote transcriptionally active euchromatic states helps in overcoming these challenges. Ultimately, understanding these epigenetic switches will be essential for unlocking the potential of important crops and advancing biotechnology to support global food security. These findings highlight that epigenetic mechanisms function as critical gatekeepers of the totipotent state by balancing repression and activation of reprogramming genes.

References

Azizi P, Hanafi MM, Sahebi M, Harikrishna JA, Taheri S, Yassoralipour A, Nasehi A (2020) Epigenetic changes and their relationship to somaclonal variation: a need to monitor the micropropagation of plantation crops. Funct Plant Biol 47(6):508–523

Baklouti E, Beulé T, Nasri A, Ben Romdhane A, Drira R, Doulbeau S, Rival A, Drira N, Fki L (2022) 2, 4-D induction of somaclonal variations in in vitro grown date palm (Phoenix dactylifera L. cv Barhee). Plant Cell Tissue Organ Cult 150(1):191–205

Bidabadi SS, Jain SM (2020) Cellular, molecular, and physiological aspects of in vitro plant regeneration. Plants 9(6):702

Bouyer D, Roudier F, Heese M, Andersen ED, Gey D, Nowack MK, Goodrich J, Renou JP, Grini PE, Colot V, Schnittger A (2011) Polycomb repressive complex 2 controls the embryo-to-seedling phase transition. PLoS Genet 7(3):e1002014

Carra A, Wijerathna-Yapa A, Pathirana R, Carimi F (2024) Development and applications of somatic embryogenesis in grapevine (Vitis spp.). Plants 13(22):3131

Cazzonelli CI, Vanstraelen M, Simon S, Yin K, Carron-Arthur A, Nisar N, Tarle G, Cuttriss AJ, Searle IR, Benkova E, Pogson BJ (2013) Role of the Arabidopsis PIN6 auxin transporter in auxin homeostasis and auxin-mediated development. PLoS One 8(7):e70069

Chang L, Zhang Z, Han B, Li H, Dai H, He P, Tian H (2009) Isolation of DNA-methyltransferase genes from strawberry (Fragaria× ananassa Duch.) and their expression in relation to micropropagation. Plant Cell Rep 28(9):1373–1384

Chanvivattana Y, Bishopp A, Schubert D, Stock C, Moon YH, Sung ZR, Goodrich J (2004) Interaction of Polycomb-group proteins controlling flowering in Arabidopsis. 131:5263

Chen X, Xu X, Shen X, Li H, Zhu C, Chen R, Munir N, Zhang Z, Chen Y, Xuhan X, Lai Z (2020) Genome-wide investigation of DNA methylation dynamics reveals a critical role of DNA demethylation during the early somatic embryogenesis of Dimocarpus longan Lour. Tree Physiol 40(12):1807–1826

Debnath SC, Ghosh A (2022) Phenotypic variation and epigenetic insight into tissue culture berry crops. Front Plant Sci 13:1042726

De-la-Peña C, Nic-Can G, Ojeda G, Herrera-Herrera JL, López-Torres A, Wrobel K, Robert-Díaz ML (2012) KNOX1 is expressed and epigenetically regulated during in vitro conditions in agave spp. BMC Plant Biol 12(1):203

Ding Y, Wang X, Su L, Zhai J, Cao S, Zhang D, Liu C, Bi Y, Qian Q, Cheng Z, Cao X (2007) SDG714, a histone H3K9 methyltransferase, is involved in Tos17 DNA methylation and transposition in rice. Plant Cell 19(1):9–22

Dong X, Gao J, Jiang M, Tao Y, Chen X, Yang X, Wang L, Jiang D, Xiao Z, Bai X, He F (2024) The identification and characterization of WOX family genes in Coffea arabica reveals their potential roles in somatic embryogenesis and the cold-stress response. Int J Mol Sci 25(23):13031

Dubois A, Roudier F (2021) Deciphering plant chromatin regulation via CRISPR/dCas9-based epigenome engineering. Epigenomes 5(3):17

Fambrini M, Usai G, Pugliesi C (2022) Induction of somatic embryogenesis in plants: different players and focus on WUSCHEL and WUS-RELATED HOMEOBOX (WOX) transcription factors. Int J Mol Sci 23(24):15950

Feng D, Liang Z, Wang Y, Yao J, Yuan Z, Hu G, Qu R, Xie S, Li D, Yang L, Gu X (2022) Chromatin accessibility illuminates single-cell regulatory dynamics of rice root tips. BMC Biol 20(1):274

Fu X, Li C, Liang Q, Zhou Y, He H, Fan LM (2016) CHD3 chromatin-remodeling factor PICKLE regulates floral transition partially via modulating LEAFY expression at the chromatin level in Arabidopsis. Sci China Life Sci 59(5):516–528

Fukai E, Umehara Y, Sato S, Endo M, Kouchi H, Hayashi M, Stougaard J, Hirochika H (2010) Derepression of the plant chromovirus LORE1 induces germline transposition in regenerated plants. PLoS Genet 6(3):e1000868

Gazzarrini S, Song L (2024) LAFL factors in seed development and phase transitions. Annu Rev Plant Biol 75(1):459–488

Grzybkowska D, Morończyk J, Wójcikowska B, Gaj MD (2018) Azacitidine (5-AzaC)-treatment and mutations in DNA methylase genes affect embryogenic response and expression of the genes that are involved in somatic embryogenesis in Arabidopsis. Plant Growth Regul 85(2):243–256

He C, Chen X, Huang H, Xu L (2012) Reprogramming of H3K27me3 is critical for acquisition of pluripotency from cultured Arabidopsis tissues. PLoS Genet 8(8):e1002911

Huang Y, Guo J, He XJ, Li C (2025) Chromatin remodeling in plants: complex composition, mechanistic diversity, and biological functions. Mol Plant 18(9):1436–1457

Hung FY, Feng YR, Hsin KT, Shih YH, Chang CH, Zhong W, Lai YC, Xu Y, Yang S, Sugimoto K, Wu K (2023) Arabidopsis histone H3 lysine 9 methyltransferases KYP/SUVH5/6 are involved in leaf development by interacting with AS1-AS2 to repress KNAT1 and KNAT2. Commun Biol 6(1):219

Ishihara H, Sugimoto K, Tarr PT, Temman H, Kadokura S, Inui Y, Sakamoto T, Sasaki T, Aida M, Suzuki T, Matsunaga S (2019) Primed histone demethylation regulates shoot regenerative competency. Nat Commun 10(1):1786

Jeevaraj T, Blessy A, Krishnamoorthy S, Sridhar A (2025) Histone demethylation by JMJ family genes: insights into plant growth and adaptation. Plant Mol Biol 115(6):1–18

Karim R, Tan YS, Singh P, Khalid N, Harikrishna JA (2018) Expression and DNA methylation of SERK, BBM, LEC2 and WUS genes in in vitro cultures of Boesenbergia rotunda (L.) Mansf. Physiol Mol Biol Plants 24(5):741–751

Kiełkowska A, Brąszewska A (2025) Demethylating drugs alter protoplast development, regeneration, and the genome stability of protoplast-derived regenerants of cabbage. BMC Plant Biol 25:463

Kowalik S, Samoń M, Przyborowski M (2026) Molecular regulators of in vitro regeneration in wheat: roles of Morphogenic factors in transformation, genome editing, and breeding. Int J Mol Sci 27(3):1271

Kumar V, Van Staden J (2017) New insights into plant somatic embryogenesis: an epigenetic view. Acta Physiol Plant 39(9):194

Lafos M, Kroll P, Hohenstatt ML, Thorpe FL, Clarenz O, Schubert D (2011) Dynamic regulation of H3K27 trimethylation during Arabidopsis differentiation. PLoS Genet 7(4):e1002040

Le H, Simmons CH, Zhong X (2025) Functions and mechanisms of histone modifications in plants. Annu Rev Plant Biol 76:551

Leduque B, Edera A, Vitte C, Quadrana L (2024) Simultaneous profiling of chromatin accessibility and DNA methylation in complete plant genomes using long-read sequencing. Nucleic Acids Res 52(11):6285–6297

Li W, Liu H, Cheng ZJ, Su YH, Han HN, Zhang Y, Zhang XS (2011) DNA methylation and histone modifications regulate de novo shoot regeneration in Arabidopsis by modulating WUSCHEL expression and auxin signaling. PLoS Genet 7(8):e1002243

Li J, Wang M, Li Y, Zhang Q, Lindsey K, Daniell H et al (2019) Multi-omics analyses reveal epigenomics basis for cotton somatic embryogenesis through successive regeneration acclimation process. Plant Biotechnol J 17(2):435–450

Liu Y, Liu K, Yin L, Yu Y, Qi J, Shen WH, Zhu J, Zhang Y, Dong A (2019) H3K4me2 functions as a repressive epigenetic mark in plants. Epigenetics Chromatin 12(1):40

Lu C, Wei Y, Abbas M, Agula H, Wang E, Meng Z, Zhang R (2024) Application of single-cell assay for transposase-accessible chromatin with high throughput sequencing in plant science: advances, technical challenges, and prospects. Int J Mol Sci 25(3):1479

Majumder S, Igamberdiev AU, Debnath SC (2025) Somaclonal variation and clonal fidelity in commercial micropropagation: challenges and perspectives. Agronomy 15(6):1489

Morinaka H, Mamiya A, Tamaki H, Iwamoto A, Suzuki T, Kawamura A, Ikeuchi M, Iwase A, Higashiyama T, Sugimoto K, Sugiyama M (2021) Transcriptome dynamics of epidermal reprogramming during direct shoot regeneration in Torenia fournieri. Plant Cell Physiol 62(8):1335–1354

Nic-Can GI, López-Torres A, Barredo-Pool F, Wrobel K, Loyola-Vargas VM, Rojas-Herrera R, De-la-Peña C (2013) New insights into somatic embryogenesis: LEAFY COTYLEDON1, BABY BOOM1 and WUSCHEL-RELATED HOMEOBOX4 are epigenetically regulated in Coffea canephora. PLoS One 8(8):e72160

Nowak K, Wójcikowska B, Gajecka M, Elżbieciak A, Morończyk J, Wójcik AM, Żemła P, Citerne S, Kiwior-Wesołowska A, Zbieszczyk J, Gaj MD (2024) The improvement of the in vitro plant regeneration in barley with the epigenetic modifier of histone acetylation, trichostatin a. J Appl Genet 65(1):13–30

Peng J, Zhang WJ, Zhang Q, Su YH, Tang LP (2023) The dynamics of chromatin states mediated by epigenetic modifications during somatic cell reprogramming. Front Cell Dev Biol 11:1097780

Rai P, Kumari P, Gaur V (2025) Erasing methylation marks on DNA by plant-specific DEMETER family DNA glycosylases. J Plant Growth Regul 44(5):1810–1826

Ramakrishnan M, Zhou M, Ceasar SA, Ali DJ, Maharajan T, Vinod KK, Sharma A, Ahmad Z, Wei Q (2023) Epigenetic modifications and miRNAs determine the transition of somatic cells into somatic embryos. Plant Cell Rep 42(12):1845–1873

Rudolf J, Tomovicova L, Panzarova K, Fajkus J, Hejatko J, Skalak J (2024) Epigenetics and plant hormone dynamics: a functional and methodological perspective. J Exp Bot 75(17): 5267–5294

Runge JS, Raab JR, Magnuson T (2016) Epigenetic regulation by ATP-dependent chromatin-remodeling enzymes: SNF-ing out crosstalk. Curr Top Dev Biol 117:1–13

Shen Q, Lin Y, Li Y, Wang G (2021) Dynamics of H3K27me3 modification on plant adaptation to environmental cues. Plants 10(6):1165

Shimooka Y, Nishikawa JI, Ohyama T (2013) Most methylation-susceptible DNA sequences in human embryonic stem cells undergo a change in conformation or flexibility upon methylation. Biochemistry 52(8):1344–1353

Shu J, Chen C, Thapa RK, Bian S, Nguyen V, Yu K, Yuan ZC, Liu J, Kohalmi SE, Li C, Cui Y (2019) Genome-wide occupancy of histone H3K27 methyltransferases CURLY LEAF and SWINGER in Arabidopsis seedlings. Plant Direct 3(1):e00100

Sivanesan I, Nayeem S, Venkidasamy B, Kuppuraj SP, Rn C, Samynathan R (2022) Genetic and epigenetic modes of the regulation of somatic embryogenesis: a review. Biologia Futura 73(3):259–277

Song YG, Liu YL, Qiu NW, Dong W (2017) Involvement of histone modification in regulating CUP-SHAPED COTYLEDON genes during shoot regeneration in Arabidopsis. Biol Plant 61(1):197–200

Suzuki N, Ochi H (2020) Regeneration enhancers: a clue to reactivation of developmental genes. Develop Growth Differ 62(5):343–354

Tanaka M, Kikuchi A, Kamada H (2008) The Arabidopsis histone deacetylases HDA6 and HDA19 contribute to the repression of embryonic properties after germination. Plant Physiol 146(1):149–161

Torres IO, Fujimori DG (2015) Functional coupling between writers, erasers and readers of histone and DNA methylation. Curr Opin Struct Biol 35:68–75

Uddenberg D, Valladares S, Abrahamsson M, Sundström JF, Sundås-Larsson A, von Arnold S (2011) Embryogenic potential and expression of embryogenesis-related genes in conifers are affected by treatment with a histone deacetylase inhibitor. Planta 234(3):527–539

Valencia-Lozano E, Cabrera-Ponce JL, Barraza A, López-Calleja AC, García-Vázquez E, Rivera-Toro DM et al (2024) Editing of SlWRKY29 by CRISPR-activation promotes somatic embryogenesis in Solanum lycopersicum cv. Micro-Tom. PLoS One 19(4):e0301169

Valledor L, Meijón M, Hasbún R, Cañal MJ, Rodríguez R (2010) Variations in DNA methylation, acetylated histone H4, and methylated histone H3 during Pinus radiata needle maturation in relation to the loss of in vitro organogenic capability. J Plant Physiol 167(5):351–357

Wada N, Ueta R, Osakabe Y, Osakabe K (2020) Precision genome editing in plants: state-of-the-art in CRISPR/Cas9-based genome engineering. BMC Plant Biol 20(1):234

Wang Y, Wang JH, Guo PR, Peng J, Xie CZ, Shi YD, Wu YH, Liao DZ, Sheng S (2025) Developmental regulatory genes in recalcitrant forest trees: advances in somatic regeneration and genetic transformation. Front Plant Sci 16:1689705

Williams L, Zhao J, Morozova N, Li Y, Avivi Y, Grafi G (2003) Chromatin reorganization accompanying cellular dedifferentiation is associated with modifications of histone H3, redistribution of HP1, and activation of E2F-target genes. Develop Dynam Off Publ American Assoc Anatom 228(1):113–120

Wójcik AM (2025) LEC2 unlocks totipotency by unlocking chromatin. Trends Plant Sci 31:129–131

Wójcikowska B, Wójcik AM, Gaj MD (2020) Epigenetic regulation of auxin-induced somatic embryogenesis in plants. Int J Mol Sci 21(7):2307

Xu P, Zhong Y, Xu A, Liu B, Zhang Y, Zhao A, Yang X, Ming M, Cao F, Fu F (2024) Application of developmental regulators for enhancing plant regeneration and genetic transformation. Plants 13(9):1272

Yadav S, Meena S, Kalwan G, Jain PK (2024) DNA methylation: an emerging paradigm of gene regulation under drought stress in plants. Mol Biol Rep 51(1):311

Yin Z, Awan MJA, Mansoor S (2025) Understanding the genetics of cotton regeneration to help improve cotton. Trends Plant Sci 31:253–256

Zhang K, Xu W, Wang C, Yi X, Zhang W, Su Z (2017) Differential deposition of H2A. Z in combination with histone modifications within related genes in Oryza sativa callus and seedling. Plant J 89(2):264–277

Zheng B, Liu J, Gao A, Chen X, Gao L, Liao L, Luo B, Ogutu CO, Han Y (2022) Epigenetic reprogramming of H3K27me3 and DNA methylation during leaf-to-callus transition in peach. Hortic Res 9:uhac132

Zilberman D, Coleman-Derr D, Ballinger T, Henikoff S (2008) Histone H2A. Z and DNA methylation are mutually antagonistic chromatin marks. Nature 456(7218):125–129

Chapter 5
Gene Regulatory Networks Orchestrating Somatic Embryogenesis

Abstract Somatic embryogenesis (SE) is a developmental process in which differentiated somatic cells regain totipotency and form embryos capable of developing into whole plants. This phenomenon serves as an important model for studying plant cell reprogramming and is widely used in plant biotechnology, genetic transformation, and large-scale propagation. SE is controlled by complex gene regulatory networks (GRNs) involving several key transcription factors, including *LEAFY COTYLEDON1* (*LEC1*), *LEAFY COTYLEDON2* (*LEC2*), *FUSCA3* (*FUS3*), *BABY BOOM* (*BBM*), *WUSCHEL* (*WUS*), and *AGAMOUS-LIKE15* (*AGL15*). These regulators coordinate with plant hormone signaling pathways, particularly auxin and cytokinin, to initiate and maintain embryogenic competence. Epigenetic mechanisms also play a vital role in regulating somatic embryogenesis. Robust changes in deoxyribonucleic nucleic acid (DNA) methylation, histone modifications, and chromatin structure contribute to the establishment of a permissive genomic environment that allows embryogenesis-related genes to be activated. Furthermore, noncoding ribose nucleic acid (RNA) contributes to the fine regulation of gene expression and hormonal signaling throughout different stages of embryo development. Comparative studies between embryogenic and non-embryogenic tissues revealed significant variations in transcriptional activity, hormonal responsiveness, and epigenetic organization. Modern technologies such as single-cell RNA sequencing, assay for transposase-accessible chromatin using sequencing (ATAC-Seq), and clustered regularly interspaced short palindromic repeats (CRISPR)-mediated genome editing provide powerful tools for reconstructing GRNs involved in SE. Understanding these regulatory processes will contribute to improved strategies for plant propagation, genetic improvement, and conservation of plant genetic resources.

Keywords Epigenetic regulation · Gene regulatory network · Somatic embryogenesis · Totipotency · Transcription factor

P. Kumar, *Totipotency in Plants*, SpringerBriefs in Plant Science,
https://doi.org/10.1007/978-981-92-1631-4_5

5.1 Introduction

Somatic embryogenesis (SE) represents a developmental transition in plants in which differentiated somatic cells are reprogrammed to initiate embryogenic pathways and form bipolar embryos in vitro without fertilization. This process exemplifies the inherent developmental plasticity of plant cells and their capacity to reacquire totipotency under appropriate inductive conditions (Fehér 2019). Unlike animal systems, where developmental potential becomes progressively restricted, plant cells retain remarkable regenerative competence that can be unlocked through hormonal, stress, or genetic cues. Consequently, SE serves not only as a practical regeneration system but also as a powerful experimental model for studying cellular reprogramming, fate determination, and developmental plasticity. From a biotechnological perspective, SE underpins clonal propagation of elite genotypes, facilitates genetic transformation and genome editing, supports synthetic seed technology, and enables long-term conservation of valuable germplasm resources (Wang et al. 2022). Unlike zygotic embryogenesis, which follows a highly structured, genetically planned developmental framework initiated by fertilization, SE requires significant reorganization of pre-existing somatic transcriptional programs. Somatic cells must suppress lineage-specific gene expression patterns and activate embryogenesis-associated regulatory modules. This transition involves coordinated shifts in hormone signaling pathways, transcription factor hierarchies, chromatin organization, epigenetic states, and metabolic profiles. Central to this transformation are GRNs, which integrate endogenous and exogenous signals into coherent transcriptional outputs that drive cell fate transitions (Wang et al. 2022). GRNs in SE are multilayered systems composed of hormone-responsive regulators; master embryogenic transcription factors such as *LEC*, *BBM*, and *WUS*; chromatin remodelers; histone modifiers; and noncoding RNAs that collectively coordinate the acquisition and stabilization of embryogenic identity. The induction of SE typically begins with exposure to high concentrations of auxin, particularly synthetic auxin such as 2,4-D, which triggers widespread transcriptional reprogramming. Auxin signaling activates *AUXIN RESPONSE FACTORs* (*ARFs*) that regulate downstream embryogenic genes and initiate feedback loops reinforcing auxin biosynthesis and signaling. This hormone-driven reprogramming is accompanied by chromatin remodeling events that increase accessibility at key regulatory loci, enabling transcription factor binding and activation of totipotency-associated genes. Epigenetic modifications including dynamic changes in DNA methylation and histone marks which further modulate transcriptional competence and ensure stable commitment to embryogenic development. Thus, SE is not a simple dedifferentiation process but rather a highly orchestrated transition toward a new, stable developmental attractor state defined by a distinct GRN configuration. Comparative transcriptomic and epigenomic analyses have revealed pronounced differences between embryogenic callus (EC) and non-embryogenic callus (NEC) tissues. Embryogenic tissues exhibit upregulation of master regulators such as *LEC2*, *BBM*, FUSCA3 (*FUS3*), and *AGL15*, along with enhanced expression of auxin biosynthesis and transport genes, cell cycle regulators and stress-responsive factors. In contrast, NEC tissues often retain expression of differentiation-associated genes, display reduced activation of embryogenic transcription factor modules, and maintain more restrictive chromatin states (Dai and Li 2024). These findings indicated that totipotency acquisition is governed not merely by the presence of hormonal stimuli but by the establishment of a self-reinforcing regulatory network architecture. In embryogenic systems, positive feedback loops among transcription factors and hormone pathways stabilize embryogenic identity, whereas in non-embryogenic systems, regulatory fragmentation and epigenetic constraints prevent network consolidation (Kumar et al. 2021). Understanding the structure and dynamics of GRNs in SE is therefore critical for elucidating the molecular basis of plant totipotency. From a fundamental perspective, dissecting these networks provides insight into how cells traverse developmental boundaries and re-establish embryonic programs. From an applied standpoint, detailed knowledge of GRN components and interactions offers opportunities to enhance regeneration efficiency, particularly in recalcitrant crops where transformation and tissue culture responses remain limiting factors. By identifying key regulatory nodes and network bottlenecks, it may become possible to engineer or chemically modulate pathways that promote embryogenic competence. This chapter brings together the current knowledge about the hormonal, transcriptional, and epigenetic frameworks that govern somatic embryogenesis. It also examines the differences in regulatory logic between embryogenic and non-embryogenic systems and discusses new tools for rebuilding and changing GRNs to make plant regeneration technologies better.

5.2 Conceptual Framework: GRN in Developmental Reprogramming

5.2.1 GRN Layout and Developmental Control

GRNs can be viewed as highly coordinated control systems that guide when and how genes are turned on or off during plant development. Instead of working as step-by-step pathways, GRNs function as interconnected networks made up of genes, transcription factors, signaling molecules, chromatin regulators, epigenetic modifiers, and noncoding RNAs. These components continuously interact with one another, allowing cells to respond flexibly to internal signals and external conditions while maintaining developmental stability. One of the most distinguishing properties of GRNs is their hierarchical nature. At the top are master transcription factors, which combine environmental and physiological stimuli to cause large-scale changes in gene expression. These master regulators activate secondary transcription factor groups and downstream genes that perform specialized developmental roles. Such organization ensures that complex developmental processes are carried out in a coordinated and precise manner. GRNs also contain regulatory loops, such as feedback and feed-forward processes, which serve to stabilize developmental decisions. Positive feedback improves cell fate commitment, whereas negative feedback reduces fluctuations and maintains gene expression balanced (Wang et al. 2020). Another key feature of GRNs is their modular structure. Gene groups frequently serve as functional modules that may be utilized at many developmental phases, such as hormone response, stress adaptability and cell cycle regulation. This modular design allows plants to adapt their developmental paths in response to environmental and physiological needs. GRNs regulate critical processes in plant development, including cellular differentiation, organ formation and embryonic patterning, by maintaining stable transcriptional states supported by certain chromatin configurations (Wang et al. 2022). However, during SE, these networks must undergo substantial reorganization. Differentiated somatic cells suppress their original vegetative programs while activating embryo-related gene networks involved in cell division, totipotency and embryonic development. Hormonal signaling, especially auxin, plays an essential role in initiating this reprogramming. Auxin-responsive transcription factors activate key embryogenic regulators such as *LEC2* and *BBM*, which subsequently initiate downstream developmental pathways. At the same time, chromatin remodeling and epigenetic changes make embryogenesis-related genes more accessible for transcription. Noncoding RNAs enhance this process by fine-tuning gene expression. Overall, a coordinated combination of hormonal signals, transcriptional regulators and epigenetic mechanisms allows somatic cells to successfully shift to an embryogenic state (Cao et al. 2019).

5.2.2 Reprogramming, Totipotency, and Network Logic

The conversion of a somatic cell into an embryogenic cell is a highly coordinated reprogramming event that can be explained using Waddington's epigenetic landscape, where different cell types exist as stable states that are maintained by GRNs. Normally, differentiated somatic cells remain in a stable vegetative state further supported by specific gene expression patterns and epigenetic marks. To gain totipotency, these cells must overcome internal regulatory barriers, reorganize their transcriptional programs, and gradually transition toward a new stable state associated with embryogenic identity. This process does not occur instantly but involves stepwise remodeling of regulatory networks (Fehér 2019). At the molecular level, SE causes large-scale changes in gene expression, chromatin structure and epigenetic control. In the early stages, stress-related and hormone-responsive genes are activated, followed by the expression of key transcription factors associated with totipotency and embryo formation (Ikeuchi et al. 2013; Wójcikowska et al. 2020). At the same time, chromatin becomes accessible at an important developmental gene, allowing transcription factors to reach and activate them. Repressive chromatin characteristics linked to somatic identity are gradually diminished, whereas changes in DNA methylation and histone marks generate a more permissive environment that promotes developmental plasticity (Kumar and Van Staden 2017; Wang et al. 2022).

Feedback interactions within GRNs are crucial for maintaining the embryogenic state once it begins. Positive feedback loops between transcription factors and hormone signaling pathways contribute to the expression of embryogenesis-related genes and ensure long-term developmental commitment. In this regard, regulators like *LEC2* and *BBM* may boost auxin signaling, enhance their activity, and promote embryogenic competence. Negative feedback systems help to maintain balance and prevent excessive dedifferentiation. Hierarchical GRN models further highlight the importance of auxin as an upstream signal that promotes chromatin remodeling and generates permissive regulatory regions. These areas enable the activation of master embryogenic transcription factors, which then activate downstream gene networks necessary for embryo development. Overall, the development of totipotency during SE is a gradual and closely regulated process that involves coordinated changes in chromatin organization, transcription factor activity, hormonal signaling, and feedback regulation within gene regulatory networks.

5.3 Stress-Induced Activation of Embryogenic GRNs

Stress signals play a crucial role in activating embryogenic GRNs during SE, serving as important factors of cellular reprogramming. Signals such as wounding, osmotic stress, temperature changes, and oxidative stress can induce complex cellular responses that help convert differentiated somatic cells into an embryonic state. These stress signals frequently interact with plant hormones, particularly auxin, to initiate the transcriptional changes required for embryogenic development (Fehér 2019; Lee et al. 2024a, 2024b). Among the various stress mediators, reactive oxygen species are particularly significant. Rather than acting only as harmful byproducts, moderate levels of reactive oxygen species function as signaling molecules that activate key transcription factors and chromatin remodeling mechanisms (Lee et al. 2024a, 2024b). Evidence has shown that reactive oxygen species signaling can occur prior to epigenetic changes such as histone acetylation and DNA methylation. These epigenetic changes relax chromatin structure, allowing important embryonic genes like *LEC1*, *WUS*, and *BBM* to be activated (Xu et al. 2024). As a result, stress-epigenetic interaction promotes a more permissive transcriptional environment, facilitating the development of cellular totipotency. Additionally, stress-responsive transcription factor families including *AP2/ERF*, *NAC*, and *WRKY* (WRKY amino acid motif) interact closely with hormone-regulated pathways to enhance embryonic GRNs. As highlighted by Fehér Attila, stress signals can function as secondary triggers, reactivating developmental programs that are generally inhibited in mature somatic cells (Fehér 2019). Recent transcriptome investigations also indicated that stress conditions increase auxin production and transport, thereby helping establish stable auxin gradients required for embryo initiation. Moreover, stress treatments have been discovered to enhance the expression of chromatin remodelers and SE receptor-like kinases (*SERKs*), which serve as early markers of the embryogenic transition. However, controlled stress has been observed to improve regeneration efficiency by activating earlier silent embryogenic pathways (Yuan et al. 2024). However, excessive stress can be harmful and may cause cell damage or death. Therefore, maintaining an ideal stress balance is critical. Overall, stress should be considered as a developmental signal that uses hormonal, transcriptional, and epigenetic pathways to promote SE and cellular plasticity.

5.4 Core Transcription Factor Module

SE is controlled by a well-organized network of transcription factors that work together to establish and maintain embryonic identities. The genes/TFs such as *LEC1*, *LEC2*, *FUS3*, and *ABSCISIC ACID INSENSITIVE3* (*ABI3*) act as the core of this network. These regulators are essential for activating seed-specific genes, promoting storage reserve accumulation, and guiding proper embryo patterning (Ikeuchi et al. 2019). As a result, it reactivates developmental programs that are normally active only during zygotic embryogenesis. Rather than functioning independently, these factors interact closely, forming a coordinated system in which each component reinforces and stabilizes the embryogenic state. Among them, *LEC2* is commonly referred to as a master regulator because its ectopic expression alone can initiate somatic embryo formation (Liu et al. 2021). This effect is linked to its ability to stimulate auxin biosynthesis and activate other embryogenesis-related transcription factors. By increasing auxin synthesis and downstream gene expression, *LEC2* creates a positive feedback loop that locks cells into an embryogenic developmental pathway and enhances their developmental flexibility. Another important regulator is *BBM*, a member of the *AP2/ERF* transcription factor family known for promoting cell proliferation and embryogenic competence. When overexpressed, *BBM* induces spontaneous somatic embryos in a variety of plant species, including *Arabidopsis*. It accomplishes this by activating genes involved in cell cycle progression and auxin pathways, resulting in the reprogramming of differentiated somatic cells into embryogenic cells. Similarly, *WUS* and other *WOX* family proteins play a key role in maintaining stem cell identity and meristem formation. During SE, increasing *WUS* expression has been strongly associated with embryogenic competence (Feng et al. 2020). It interacts with cytokinin signaling pathways, which help define embryonic meristematic regions and promote prolonged regenerative growth. Furthermore, the *MADS-box* transcription factor *AGAMOUS-LIKE15* (*AGL15*) promotes embryonic development by regulating hormone-responsive genes and coordinating auxin and ethylene signaling. *AGL15* helps to maintain the stability of embryogenic transcriptional program by maintaining the expression of essential developmental regulators. The *LEC* module, *BBM*, *WUS*, and *AGL15* forms an integrated regulatory network that combines hormonal signaling, transcriptional control, and epigenetic regulation, allowing somatic cells to enter and remain in an embryogenic state throughout plant regeneration (Xu et al. 2024).

5.5 Epigenetic Regulation of Embryogenic GRNs

5.5.1 DNA Methylation

DNA methylation is a key epigenetic mechanism that influences gene expression by modifying cytosine residues in the genome, which often leads to transcriptional repression when present in gene promoters or regulatory regions. During SE, embryogenic tissues frequently show lower levels of DNA methylation compared to non-embryogenic callus. It has been reported that a decrease in methylation enables the activation of developmental programs required for embryo formation by loosening chromatin structure and increasing the accessibility of previously silenced embryogenesis-related loci to transcription factors and other regulatory proteins (Mozgova and Hennig 2015). Recent genome-wide analyses have reinforced the idea that dynamic changes in DNA methylation are tightly linked to totipotency acquisition. For example, several comparative methylome studies in species such as *Arabidopsis* have shown that embryogenic callus or tissues undergoing SE exhibit targeted demethylation at key developmental regulators including *LEC1*, *LEC2*, *BBM*, and *WUS*, coinciding with their elevated expression (Lee et al. 2024a, 2024b). These demethylation events often occur in both cytosine-guanine (CG) and non-CG sequence contexts and are linked with increased chromatin accessibility, suggesting coordinated epigenetic resetting during the embryogenic transition. Loss of methylation in embryogenic tissues has been frequently accompanied by the downregulation of methyltransferase genes (e.g., *MET1*, Chromomethylase 3 (*CMT3*)) and upregulation of demethylation-related genes such as *ROS1* (*Repressor Of Silencing 1*) and *DML* (*DEMETER-like*) family members, indicating active regulation of the methylome rather than passive dilution (Ikeuchi et al. 2019). This regulatory change appears to be both locus-specific and developmental in nature, with early stages of SE showing the most significant methylation loss at embryogenesis-specific promoters and enhancers. In addition to its direct effects on developmental regulators, reduced DNA methylation also impacts hormone signaling pathways, stress response genes, and small RNA biogenesis components, further influencing the capacity of cells to respond to inductive signals like auxin (Yang et al. 2024). Studies using methylation inhibitors such as 5-azacytidine have demonstrated that artificially reducing DNA methylation can enhance embryogenic competence, highlighting the functional importance of methylation dynamics in SE (Gantait and Chmielarz 2025).

5.5.2 Histone Modifications

Histone modifications play a central role in regulating chromatin dynamics during SE and the development of totipotency. Among these, the repressive histone mark *H3K27me3* (trimethylation of lysine 27 on histone H3) is particularly important because it maintains developmental genes in a silenced state under normal conditions. During SE induction, extensive chromatin remodeling occurs, leading to the removal or redistribution of *H3K27me3* marks from key embryogenic regulators such as *LEC1*, *LEC2*, *WUS*, and *BBM*, thereby enabling their transcriptional activation and promoting embryogenic competence (Yang et al. 2024). Polycomb Repressive Complex 2 (*PRC2*) is the key epigenetic regulator responsible for depositing *H3K27me3* marks and maintaining stable gene repression during plant development. Vijayanathan et al. (2022) suggest that modulation of *PRC2* activity is essential for cellular reprogramming and for maintaining totipotency. Reduced *PRC2*-mediated repression has been associated with enhanced chromatin accessibility and activation of embryogenesis-related transcription factors during callus formation and SE in model plants such as *Arabidopsis*. This epigenetic relaxation facilitates the shift from a differentiated to an embryogenic state by resetting developmental gene expression programs. In addition to *H3K27me3* removal, activating histone marks such as *H3K4me3* and histone acetylation (e.g., *H3K9ac and H3K27ac*) are enriched at embryogenic loci during SE induction (Mozgova and Hennig 2015). These activating marks promote open chromatin structure and enhance transcription of genes involved in hormone signaling, cell division, and developmental plasticity. Histone acetyltransferases (HATs) and histone deacetylases (HDACs) dynamically regulate this balance, and chemical inhibition of HDACs has been demonstrated to significantly improve SE efficiency in several plant systems.

5.5.3 Chromatin Remodeling

Chromatin remodeling is a fundamental epigenetic mechanism that facilitates the transition of somatic cells into an embryogenic state by dynamically altering chromatin structure and accessibility. During SE, large-scale chromatin reorganization typically occurs before the transcriptional activation of key embryogenesis-related genes, indicating that epigenetic priming comes before developmental reprogramming. Genome-wide chromatin accessibility studies, particularly ATAC-Seq and DNase I hypersensitivity analyses, have revealed increased accessibility at promoters and regulatory regions of core embryogenic regulators such as *LEC1*, *LEC2*, *BBM*, and *WUS* during early SE induction, reflecting a shift from a closed to a more permissive chromatin state (Ikeuchi et al. 2019; Han et al. 2022). This chromatin opening allows transcription factors and hormone-responsive regulators to bind to target loci more efficiently, thereby activating embryogenic gene regulatory networks. Further studies highlighted that chromatin remodeling complexes, such as *SWI/SNF* (Switch/Sucrose Non-Fermentable), *BAF* (BRG1/BRM-Associated Factors), and *CHD* (Chromodomain Helicase DNA-binding) families, play a crucial role in restructuring nucleosome positioning during embryogenic reprogramming. These ATP-dependent remodelers actively reposition or evict nucleosomes, providing access to previously silenced developmental genes and enhancing transcriptional plasticity. In this regard, *BRM* (BRAHMA) and *SYD* (SPLAYED), key *SWI/*

SNF components in plants, have been shown to regulate meristem and embryogenesis-related pathways by influencing chromatin accessibility at developmental loci. Furthermore, the interaction between chromatin remodelers and histone modification machinery creates a coordinated regulatory system, where histone acetylation and reduced repressive marks promote a more relaxed chromatin configuration conducive to embryogenic competence (Ikeuchi et al. 2019). Emerging single-cell epigenomic studies have also shown that chromatin accessibility changes are highly variable and cell-type specific during SE, with embryogenic cells having a unique open chromatin landscape compared to non-embryogenic callus. This shows that chromatin remodeling is not just a product of transcription but also an early indicator of cell fate transition. Moreover, stress signals and auxin treatments, which are used for SE induction, have been found to trigger rapid chromatin decondensation and activation of pioneer transcription factors, resulting in developmental reprogramming. Overall, chromatin remodeling acts as an upstream regulatory layer, integrating hormonal, transcriptional, and epigenetic signals, enabling the activation of embryogenic genes and the acquisition of totipotency during SE (Fehér 2019; Wang et al. 2022; Liu et al. 2024).

5.5.4 Noncoding RNAs (ncRNAs)

ncRNAs, particularly microRNAs (miRNAs), play an important role in regulating gene expression during SE. Rather than coding for proteins, these small RNA molecules regulate gene activity after transcription by directing the cleavage of target mRNAs or blocking their translation. This technique allows miRNAs to precisely control the levels of transcription factors and hormone signaling components that determine whether a cell can acquire embryogenic competence. For example, *miR160*, *miR167*, and *miR390* are well known for targeting *ARF* genes, thereby adjusting auxin signaling during SE induction (Wójcikowska et al. 2020). By altering auxin sensitivity, these miRNAs help to ensure that the key embryogenic regulators such as *LEC2*, *BBM*, and *WUS* are activated at the right time and in the correct developmental condition. The dynamic and stage-dependent miRNA expression patterns during SE in *Arabidopsis* show the high-throughput small RNA sequencing. For example, *miR156* and *miR172* coordinate developmental phase transitions that affect embryogenic potential, whereas *miR396* regulates GRF transcription factors involved in cell proliferation and callus formation (Asghar et al. 2023). Likewise, *miR393* controls auxin receptor genes (*TIR1/AFB* family), altering hormone sensitivity during early embryogenic induction. These findings highlighted how miRNAs act as buffering agents, preventing excessive or premature activation of developmental pathways. Long noncoding RNAs (lncRNAs) and small interfering RNAs (siRNAs) have also been reported for their involvement in chromatin regulation and epigenetic stability during SE. Certain lncRNAs interact with chromatin modifiers to modify histone marks at embryogenesis-related loci, linking ncRNA activity to epigenetic remodeling (Cordeiro et al. 2022). Moreover,

RNA-directed DNA methylation (RdDM), mediated by siRNAs, contributes to selective gene silencing and genome stability during the reprogramming phase. In long-term cultured callus of *Solanum betaceum*, *miR827* has been associated with the gradual loss of embryogenic competence by modulating phosphate homeostasis and sugar signaling pathways, demonstrating how metabolic and nutrient signals intersect with miRNA networks to influence SE outcomes (Cordeiro et al. 2025). Overall, noncoding RNAs provide an additional regulatory mechanism that integrates hormonal, metabolic, and epigenetic signals, ensuring precise temporal and spatial control of embryogenic gene expression and increasing the resilience of SE.

5.6 Comparative Regulatory Analysis Between Embryogenic and Non-embryogenic Systems

Embryogenic callus can be considered a highly flexible and responsive cellular state. Its open chromatin organization, reduced epigenetic barriers, and highly active hormone signaling pathways allow genes required for embryo formation to be activated more efficiently. The regulatory networks in these cells are interconnected, supporting each other to stabilize and maintain the embryogenic program. On the other hand, non-embryogenic callus exists in a more confined and developmentally fixed condition. The chromatin remains relatively compact, key developmental genes stay epigenetically silenced, and regulatory networks are less coordinated. As a result, these cells continue following a vegetative or differentiated pathway and lack the molecular context necessary to regain totipotency. The key regulatory differences between embryogenic and non-embryogenic callus are summarized in Table 5.1.

5.7 Dynamic Regulation of Totipotent Transition

The transition to totipotency during SE is controlled by a complex and dynamic regulatory system rather than a simple step-by-step pathway. During this process, differentiated plant cells gradually lose their original identity and develop the ability to generate a complete embryo. This developmental change is governed by interconnected regulatory circuits that act like bistable switches, enabling cells to move from a vegetative state to an embryogenic state when specific molecular conditions are met. When these regulatory thresholds are reached, the embryogenic program becomes stabilized, and the cells commit to embryo formation. The auxin-dependent activation of key embryogenic transcription factors, particularly *LEC2* and *BBM*, is central to this transformation. Elevated auxin levels induce the expression of these regulators, which then activate a wide range of downstream genes involved in embryo development and cellular reprogramming. Importantly, *LEC2* and *BBM*

Table 5.1 Key regulatory differences between embryogenic and non-embryogenic callus

Features	Embryogenic callus	Non-embryogenic callus	References
Transcriptomic profile	High expression of totipotency-associated genes and TFs such as *LEC1*, *LEC2*, *BBM*, *FUS3*, *AGL15*, and *WUS*	Reduced or negligible expression of core embryogenic TFs	Kumar and Van Staden (2017); Dai and Li (2024)
Gene regulatory network (GRN)	Highly interconnected network architecture. Presence of positive feedback loops among *LEC-BBM-WUS* modules	Fragmented regulatory networks with weak inter-module connectivity. Absence of stable feedback loops sustaining embryogenic identity	Ikeuchi et al. (2019); Singh et al. (2024)
Hormonal sensitivity and signaling	Elevated auxin sensitivity; enhanced expression of *PIN* transporters. Balanced cytokinin-auxin crosstalk. Reduced gibberellin signaling	Altered (or) reduced auxin responsiveness. Higher (or) unregulated gibberellin signaling	Fehér (2019); Wang et al. (2025)
DNA methylation	Reduced CG and non-CG methylation at *LEC*, *BBM*, *WUS*	Higher DNA methylation levels at embryogenic loci	Wójcikowska et al. (2020); Vijayanathan et al. (2022)
Histone modification	Reduced repressive mark- *H3K27me3* at key embryogenic genes. Increased activating marks- *H3K4me3*	Persistent *H3K27me3* deposition at developmental loci	Mozgova and Hennig (2015); Ikeuchi et al. (2019)
Chromatin accessibility	Open chromatin state revealed by ATAC-Seq; enhanced accessibility at promoters of *LEC* and *BBM*	Compact chromatin at embryogenic loci	Wang et al. (2020); Peng et al. (2023)
Noncoding RNA regulation	Dynamic expression of miRNAs (*miR160*, *miR167*, *miR390*, *miR156*, *miR396*)	Dysregulated or imbalanced miRNA expression (*miR827*)	Wójcikowska et al. (2020)
Stress response	Controlled ROS accumulation and activation of stress-responsive TFs (*AP2/ERF*, *NAC*, *WRKY*) enhancing embryogenic competence	Either weak stress signaling activation or excessive stress leading to cell damage rather than reprogramming	Fehér (2019)

AGL15 AGAMOUS-LIKE15, *AP2/ERF* APETALA2/Ethylene-Responsive Factor, *ATAC-Seq* Assay for Transposase-Accessible Chromatin using sequencing, *BBM* BABY BOOM, CG Cytosine-Guanine, DNA deoxyribonucleic nucleic acid, *FUS3* FUSCA3, *H3K4me3* Histone H3 lysine 4 trimethylation, *LEC* LEAFY COTYLEDON, *LEC1* LEAFY COTYLEDON1, *LEC2* LEAFY COTYLEDON2, miRNA MicroRNA, *PIN* PIN-FORMED, RNA ribonucleic acid, TFs transcription factors, *H3K27me3* trimethylation of lysine 27 on histone H3, *WRKY* WRKY amino acid motif, *WUS* WUSCHEL

participate in positive feedback loops that encourage their own expression and enhance the activity of other embryogenesis-related regulators. These self-reinforcing circuits create a stable transcriptional environment that promotes the acquisition and maintenance of embryogenic competence (Su et al. 2009; Ikeuchi et al. 2019; Wang et al. 2022). These feedback-driven networks cause threshold-dependent responses in plant cells. Despite hormone stimulation, cells in the early stages remain callus or vegetative. When the combined effects of hormones, stress signals, and transcription factors reach a critical level, the regulatory network reorganizes and drives the cell into an embryogenic attractor state. This stable state represents a developmental configuration associated with totipotency and continued embryogenic growth (Fehér 2019; Wang et al. 2022). In addition to transcription factor regulation, significant crosstalk between hormonal signaling pathways, epigenetic modifiers, and gene regulatory networks helps to stabilize the embryogenic state. Chromatin remodeling and histone modifications increase the accessibility of embryogenesis-related genes, whereas hormone signaling pathways, particularly auxin and cytokinin, continue to promote developmental reprogramming. These coordinated regulatory layers ensure that once embryogenic competence is established, the developmental program remains stable and productive (Ikeuchi et al. 2019; Liu et al. 2024). The acquisition of totipotency during SE is the result of an integrated interaction of transcriptional feedback loops, hormonal signaling, and epigenetic regulation, which together create a strong molecular framework that drives and maintains embryogenic development.

5.8 Emerging Approaches for GRN Reconstruction in SE

Recent advances in molecular and computational technologies have significantly enhanced our ability to understand and reconstruct GRNs involved in SE. These modern tools provide deeper insights into the molecular events that cause cellular reprogramming, allowing researchers to examine developmental changes with greater accuracy. As a result, it is now possible to better understand the complex interactions among genes, transcription factors, hormones, and epigenetic regulators that govern the transition from a somatic cell to an embryogenic state (Fehér 2019; Ikeuchi et al. 2019). One of the most efficient techniques in this field is single-cell RNA sequencing (scRNA-Seq). Unlike conventional transcriptomic analyses that evaluate average gene expression from large groups of cells, single-cell approaches allow scientists to study gene activity in individual cells. This makes it possible to detect cellular diversity within callus tissues as well as trace how specific cells gradually progress toward an embryogenic fate. Mapping these developmental pathways allows researchers to identify intermediate cellular states and point out the regulatory genes that initiate and guide the transition to embryogenic competence (Ikeuchi et al. 2019; Zhang et al. 2025). Another approach is chromatin accessibility analysis, particularly through ATAC-Seq. This method identifies regions of the genome where chromatin is open and accessible for transcription factor binding.

ATAC-Seq studies have shown that during SE, essential developmental genes such as *LEC*, *BBM*, and *WUS* become more accessible as cells begin to acquire embryogenic potential. These findings highlight how epigenetic remodeling plays a key role in activating the transcriptional programs required for embryo formation (Han et al. 2022; Liu et al. 2024). The functional roles of genes within these regulatory networks can be directly tested using CRISPR-based genome editing technologies. By selectively modifying transcription factors, chromatin regulators or hormone-related genes, it can evaluate how individual components contribute to the regulatory circuits controlling SE. Such targeted perturbation experiments provide strong evidence for causal relationships within gene regulatory networks rather than relying solely on correlations. Overall, combining single-cell transcriptomics, chromatin accessibility profiling, and CRISPR-based functional validation offers a powerful strategy for reconstructing gene regulatory networks involved in SE. When integrated with multi-omics datasets and advanced network inference approaches, these technologies are expected to significantly advance our understanding of the molecular mechanisms that enable plant cells to undergo reprogramming and acquire totipotency.

5.9 Conclusion

Somatic embryogenesis represents a unique example of plant cellular plasticity, where differentiated somatic cells regain totipotency and develop into complete embryos. Over the past decades, substantial progress has been made in understanding the molecular and regulatory mechanisms that govern this complex developmental transition. Interconnected GRNs composed of key transcription factors such as *LEC1*, *LEC2*, *FUS3*, *BBM*, *WUS*, and *AGL15* play a crucial role in establishing and maintaining embryogenic identity. These regulators interact closely with plant hormone signaling pathways, especially auxin and cytokinin, to initiate and stabilize embryogenic competence. In addition to transcriptional regulation, epigenetic mechanisms play a crucial role in enabling the transition to totipotency. Comparative analyses between embryogenic and non-embryogenic tissues have revealed substantial differences in transcriptional activity, hormonal sensitivity, and epigenetic landscapes. Embryogenic tissues are often characterized by highly interconnected regulatory networks, enhanced responsiveness to developmental signals, and open chromatin configurations that promote gene activation. In contrast, non-embryogenic tissues often maintain more restricted regulatory states characterized by stable epigenetic repression and limited network connectivity. Recent advances in high-throughput and computational technologies are further transforming the understanding of SE regulatory mechanisms. Tools such as single-cell transcriptomics, chromatin accessibility profiling (ATAC-Seq), and CRISPR-mediated functional genomics are enabling more precise reconstruction of GRNs involved in embryogenic transitions. These approaches provide unparalleled insights into the dynamic regulatory logic underlying plant cell reprogramming and totipotency. Integrating transcriptional, epigenetic, hormonal, and computational perspectives offers a comprehensive framework for understanding SE.

References

Asghar S, Ghori N, Hyat F, Li Y, Chen C (2023) Use of auxin and cytokinin for somatic embryogenesis in plant: a story from competence towards completion. Plant Growth Regul 99(3):413–428

Cao X, Yang H, Shang C, Ma S, Liu L, Cheng J (2019) The roles of auxin biosynthesis YUCCA gene family in plants. Int J Mol Sci 20(24):6343

Cordeiro D, Canhoto J, Correia S (2022) Regulatory non-coding RNAs: emerging roles during plant cell reprogramming and in vitro regeneration. Front Plant Sci 13:1049631

Cordeiro D, Canhoto J, Correia S (2025) Analysing the loss of embryogenic competence in long-term cell lines of Solanum betaceum Cav.: involvement of miR827, phosphate and sugar. BMC Plant Biol 25(1):851

Dai L, Li T (2024) Comparative transcriptomic insights into the mechanisms underlying maize (Zea mays L.) embryogenic callus differentiation. Agronomy 14(8):1689

Fehér A (2019) Callus, dedifferentiation, totipotency, somatic embryogenesis: what these terms mean in the era of molecular plant biology? Front Plant Sci 10:536

Feng K, Hou XL, Xing GM, Liu JX, Duan AQ, Xu ZS, Li MY, Zhuang J, Xiong AS (2020) Advances in AP2/ERF super-family transcription factors in plant. Crit Rev Biotechnol 40(6):750–776

Gantait S, Chmielarz P (2025) Conservation of plant genetic resources

Han J, Lopez-Arredondo D, Yu G, Wang Y, Wang B, Wall SB, Zhang X, Fang H, Barragán-Rosillo AC, Pan X, Jiang Y, Wang K (2022) Genome-wide chromatin accessibility analysis unveils open chromatin convergent evolution during polyploidization in cotton. Proc Natl Acad Sci 119(44):e2209743119

Ikeuchi M, Sugimoto K, Iwase A (2013) Plant callus: mechanisms of induction and repression. Plant Cell 25(9):3159–3173

Ikeuchi M, Favero DS, Sakamoto Y, Iwase A, Coleman D, Rymen B, Sugimoto K (2019) Molecular mechanisms of plant regeneration. Annu Rev Plant Biol 70(1):377–406

Kumar V, Van Staden J (2017) New insights into plant somatic embryogenesis: an epigenetic view. Acta Physiol Plant 39(9):194

Kumar S, Ruggles A, Logan S, Mazarakis A, Tyson T, Bates M, Grosse C, Reed D, Li Z, Grimwood J, Schmutz J, Saski C (2021) Comparative transcriptomics of non-embryogenic and embryogenic callus in semi-recalcitrant and non-recalcitrant upland cotton lines. Plants 10(9):1775

Lee S, Bae SH, Jeon Y, Seo PJ, Choi Y (2024a) DEMETER DNA demethylase reshapes the global DNA methylation landscape and controls cell identity transition during plant regeneration. BMC Genomics 25(1):1234

Lee S, Park YS, Rhee JH, Chu H, Frost JM, Choi Y (2024b) Insights into plant regeneration: cellular pathways and DNA methylation dynamics. Plant Cell Rep 43(5):120

Liu B, Sun G, Liu C, Liu S (2021) LEAFY COTYLEDON 2: a regulatory factor of plant growth and seed development. Genes 12(12):1896

Liu J, Ke M, Sun Y, Niu S, Zhang W, Li Y (2024) Epigenetic regulation and epigenetic memory resetting during plant rejuvenation. J Exp Bot 75(3):733–745

Mozgova I, Hennig L (2015) The polycomb group protein regulatory network. Annu Rev Plant Biol 66:269–296

Peng J, Zhang WJ, Zhang Q, Su YH, Tang LP (2023) The dynamics of chromatin states mediated by epigenetic modifications during somatic cell reprogramming. Front Cell Dev Biol 11:1097780

Singh S, Praveen A, Dudha N, Sharma VK, Bhadrecha P (2024) Single-cell transcriptomics: a new frontier in plant biotechnology research. Plant Cell Rep 43(12):294

Su YH, Zhao XY, Liu YB, Zhang CL, O'Neill SD, Zhang XS (2009) Auxin-induced WUS expression is essential for embryonic stem cell renewal during somatic embryogenesis in Arabidopsis. Plant J 59(3):448–460

Vijayanathan M, Trejo-Arellano MG, Mozgová I (2022) Polycomb repressive complex 2 in eukaryotes—an evolutionary perspective. Epigenomes 6(1):3

Wang FX, Shang GD, Wu LY, Xu ZG, Zhao XY, Wang JW (2020) Chromatin accessibility dynamics and a hierarchical transcriptional regulatory network structure for plant somatic embryogenesis. Dev Cell 54(6):742–757

Wang FX, Shang GD, Wang JW (2022) Towards a hierarchical gene regulatory network underlying somatic embryogenesis. Trends Plant Sci 27(12):1209–1217

Wang Y, Wang JH, Guo PR, Peng J, Xie CZ, Shi YD, Wu YH, Liao DZ, Sheng S (2025) Developmental regulatory genes in recalcitrant forest trees: advances in somatic regeneration and genetic transformation. Front Plant Sci 16:1689705

Wójcikowska B, Wójcik AM, Gaj MD (2020) Epigenetic regulation of auxin-induced somatic embryogenesis in plants. Int J Mol Sci 21(7):2307

Xu P, Zhong Y, Xu A, Liu B, Zhang Y, Zhao A, Yang X, Ming M, Cao F, Fu F (2024) Application of developmental regulators for enhancing plant regeneration and genetic transformation. Plants 13(9):1272

Yang F, Liu X, Qiao Y, Tang X, Luo P (2024) Microspore embryogenesis: in vitro cultivation induced cell reprogramming for plant breeding. Seed Biology 3(1):0

Yuan HY, Kagale S, Ferrie AM (2024) Multifaceted roles of transcription factors during plant embryogenesis. Front Plant Sci 14:1322728

Zhang L, Wang F, Luo B, Chen N, Wang Y, Zhang X (2025) Mechanism and application of developmental factors in plant genetic transformation. Int J Mol Sci 26(20):10135

Chapter 6
Evolutionary and Species-Specific Regulation of Totipotency

Abstract The induction of a regenerative state in differentiated somatic cells through the targeted activation of transcriptional networks governing cellular totipotency has emerged as a key strategy for overcoming regeneration limitations in plants. This process is regulated by a conserved set of developmental regulatory genes, including members of the *WUSCHEL* (*WUS*), *WUSCHEL-RELATED HOMEOBOX* (*WOX*), *BABY BOOM* (*BBM*), *LEAFY COTYLEDON* (*LEC*), *SOMATIC EMBRYOGENESIS RECEPTOR-LIKE KINASE* (*SERK*), and *AGAMOUS-LIKE* (*AGL*) gene families. Although these regulators are broadly conserved across plant lineages, their activity exhibits species-specific variation due to differences in gene regulatory networks, chromatin organization, and hormonal responsiveness. This chapter examines the evolutionary conservation and divergence of key totipotency regulators across monocots, dicots, gymnosperms, and woody perennials. Comparative analyses highlight how lineage-specific modifications in transcriptional regulation and hormone-mediated signaling pathways contribute to variation in regenerative competence and explain the widespread occurrence of regeneration recalcitrance in many plant species. Particularly in recalcitrant spp.; it has limited regeneration capacity to form somatic embryogenesis and for these spp. genetic transformation remains major barriers for biotechnology applications. The chapter further discusses emerging synthetic and genetic engineering approaches, including the manipulation of developmental regulatory genes, transient expression systems, stable transformation platforms, and genome editing technologies aimed at overcoming species-specific regeneration constraints. By integrating insights from evolutionary developmental biology with modern plant biotechnology, this chapter provides a conceptual framework for understanding interspecific variation in totipotency and highlights strategies for improving regeneration systems in diverse plant species.

Keywords Developmental regulatory genes · Evolutionary · Recalcitrant · Species-specific regulation · Synthetic approaches · Totipotency

P. Kumar, *Totipotency in Plants*, SpringerBriefs in Plant Science,
https://doi.org/10.1007/978-981-92-1631-4_6

6.1 Introduction

Reprogramming differentiated somatic cells into a regenerative state through the targeted activation of transcriptional networks that govern cellular totipotency has emerged as a promising strategy for overcoming regeneration bottlenecks in plants. This regenerative ability is a highly regulated developmental program controlled by a network of key regulatory genes. At the core of this regenerative process are developmental regulatory genes, including members of the *WUSCHEL* (*WUS*), *WUSCHEL-RELATED HOMEOBOX* (*WOX*), *BABY BOOM* (*BBM*), *LEAFY COTYLEDON* (*LEC*), and *AGAMOUS-LIKE* (*AGL*) gene families (Fambrini et al. 2022). These transcriptional regulators act as key molecular determinants that orchestrate cell fate reprogramming, acquisition of embryogenic competence, and the initiation of organogenic pathways. By modulating complex gene regulatory networks, they integrate hormonal cues and epigenetic signals to direct the transition from a differentiated cellular state toward a totipotent and developmentally plastic condition (Xiao et al. 2017). Although these genes are widely conserved across plant lineages, still their functions are not similar. Instead, they exhibit species-specific regulatory behavior, functioning differently in monocots, dicots, gymnosperms, and woody perennials depending on their genomes, promoter architecture, hormonal regulation, and epigenetic landscape (Wang et al. 2025). Totipotency can be regarded as an ancestral developmental attribute that has been preserved throughout plant evolution, yet selectively refined in different lineages. Although the protein-coding regions of key developmental regulatory genes remain highly conserved, evolutionary modifications in chromatin architecture and gene regulatory network interactions have contributed to marked interspecific differences in regenerative performance (Marand et al. 2023). Comparative studies revealed that model dicots such as *Arabidopsis thaliana* and tobacco often display high embryogenic responsiveness, whereas many cereals, including rice, maize, and wheat, as well as numerous woody crops, show strong genotype dependence and recalcitrance (Wang et al. 2025). These differences indicated that evolutionary changes in regulatory organization rather than simply the presence or absence of essential developmental genes play a decisive role in determining species-specific regeneration capacity. In other words, how genes are regulated is often more important than whether they are present. Accordingly, this chapter explores the evolutionary conservation and species-specific divergence of developmental regulatory genes governing totipotency and examines how changes in gene regulatory networks and hormone-responsive pathways shape regenerative competence across plant lineages. Further, this chapter highlights emerging synthetic and genetic engineering strategies aimed at overcoming regeneration recalcitrance in diverse plant species.

6.2 Comparative Analysis of Developmental Regulatory Gene Orthologs

6.2.1 *Conserved Master Regulators of Totipotency*

Comparative analysis across angiosperms and gymnosperms revealed that a conserved set of developmental regulatory genes functions as master regulators of totipotency. Members of the *LEC* (*LEC1/LEC2*), *BBM*, *WUS/WOX*, *SOMATIC EMBRYOGENESIS RECEPTOR-LIKE KINASE* (*SERK*), and *AGL* gene families repeatedly appear as positive drivers of somatic embryogenesis, embryogenic callus formation, and shoot regeneration across phylogenetically distant taxa. Across a vast phylogenetic distance from bryophytes like *Physcomitrium patens* to complex angiosperms, these gene families act as universal switches for totipotency. In *Arabidopsis thaliana*, ectopic expression or overexpression of *AtWUS*, *AtBBM*, *AtLEC1/2*, and *AtAGL15* induces embryogenic competence, spontaneous somatic embryogenesis, and callus formation (Ledwoń and Gaj 2009; Horstman et al. 2017). Similar functional conservation has also been evident in crops. In *Oryza sativa*, overexpression of *OsSERK1* and *WOX* orthologs enhances callus initiation and shoot regeneration (Hu et al. 2017). In *Zea mays*, *ZmBBM* and *ZmWUS* co-expression markedly improves somatic embryo formation and transformation efficiency (Mookkan et al. 2017; Jiang et al. 2024). Similarly, in *Glycine max* and *Brassica napus*, *BBM* and *AGL* orthologs promote embryogenic transitions (Belide et al. 2013). Additionally, in gymnosperms such as *Picea abies* and *Pinus radiata*, *PaWOX2/8/9* and related *WOX* orthologs are expressed during embryogenic stages and promote pro-embryogenic mass formation (Hedman et al. 2013). In *Larix decidua*, *LdBBM and LdLEC1* are expressed during early embryogenesis (Rupps et al. 2016). These plant species indicated that deeply conserved regulatory gene families, mainly *LEC*, *BBM*, *WOX*/*WUS*, and *SERK*, constitute an ancestral core totipotency network retained across seed plants.

6.2.2 *Monocot-Dicot Divergence*

Despite conservation of core regulators, regeneration outcomes reveal clear monocot-dicot divergence in regulatory organization. In dicots such as *Arabidopsis*, *Nicotiana tabacum*, and *Gossypium hirsutum*, ectopic expression of *WUS*, *BBM*, and *LEC* genes often directly triggers somatic embryogenesis or shoot organogenesis without strict dependence on external hormonal cues (Guo et al. 2013; Hu et al. 2017). For example, *AtBBM* induces spontaneous somatic embryo formation in *Arabidopsis*, while *GhLEC1*/*2* promotes high-frequency embryogenic callus formation in cotton (Zheng et al. 2014). In contrast, in monocots such as maize and rice, overexpression frequently enhances regeneration efficiency rather than autonomously initiating embryogenesis. In maize, *ZmBBM/ZmWUS* increases callus induction rates and transformation success, but induction still depends on growth hormones (Mookkan et al. 2017; Jiang et al. 2024). Similarly, in rice, *OsSERK1* highlights specialized pathways for shoot regeneration and callus initiation that are distinct from the typical *AtWIND1* (Wound-Induced Dedifferentiation)-mediated pathways seen in dicots (Hu et al. 2005). In wheat (*Triticum aestivum*), overexpression of regeneration-related genes such as *TaLAX1* (*Triticum aestivum*-like Auxin Resistant 1) and *TaGRF* (*Triticum aestivum* Growth-Regulating Factor) family members improves regeneration efficiency rather than trigger spontaneous somatic embryogenesis (Yu et al. 2024). Thus, monocots often exhibit quantitative enhancement of regeneration via developmental regulatory gene manipulation, whereas dicots more frequently display qualitative fate conversion upon ectopic expression, suggesting lineage-specific regulatory sensitivity.

6.3 Expression Divergence and Network Rewiring

Comparative analyses across diverse species indicate that, although core developmental regulatory genes governing totipotency are evolutionarily conserved, their transcriptional expression is highly species- and stage-specific. Such variation reflects expression divergence and context-dependent rewiring of gene regulatory networks, often mediated by plant growth hormones. Growth hormones not only activate conserved embryogenic genes but also reshape network connectivity, alter promoter responsiveness, and modulate chromatin accessibility, thereby generating lineage-specific regeneration outcomes as shown in Table 6.1. In *Medicago truncatula, WOX, BBM*-like, and *FUSCA3* (*FUS3*)-like genes have shown stage-specific expression during heart and torpedo stages of somatic embryogenesis, with differential activation between embryogenic and non-embryogenic lines (Orłowska et al. 2017; Orłowska and Kępczyńska 2018). Hormonal modulation further reinforces this divergence: exogenous gibberellic acid (GA_3) suppresses callus and somatic embryo formation by downregulating *MtBBM1*, whereas paclobutrazol, a gibberellin biosynthesis inhibitor, enhances *MtBBM1* expression without impairing embryogenesis (Igielski and Kępczyńska 2017). Additionally, GA-mediated regulation of Copalyl Diphosphate Synthase (*MtCPS*), Gibberellin 3-Oxidase 1 (*MtGA$_3$OX1*) and Gibberellin 3-Oxidase 2 (*MtGA$_3$OX2*) modifies endogenous bioactive GA levels during embryo development, illustrating hormone-driven rewiring of the embryogenic network. Auxin-cytokinin combinations (NAA (1-Naphthaleneacetic Acid) + BAP (6-Benzylaminopurine) + IAA (Indole-3-Acetic Acid) further induce *MtWUS, MtSERK1, MtWOX11-like*, and *MtPIN10* (PIN-FORMED 10), linking phytohormone signaling interplay to stem cell specification and organogenic competence (Tvorogova et al. 2015).

In citrus (*Citrus sinensis*), overexpression of *CsFUS3* enhances embryogenic callus formation and somatic embryo induction, whereas mutation of this gene perturbs embryogenic progression (Zhu et al. 2014). These findings suggested that hormone-responsive promoter contexts modulate *FUS3*-mediated regulatory

Table 6.1 Hormone-driven expression divergence and gene regulatory network rewiring during totipotency across plant species

Growth hormones	Species	Developmental context	Target developmental regulators	Transcriptional response	Network-level interpretation	References
2,4-D	*Arabidopsis thaliana*	Induction of somatic embryogenesis	*AtAHL15, 19, 20, 29*	Upregulated	Auxin-triggered chromatin remodeling factors facilitate reprogramming	Karami et al. (2021)
	Ananas comosus	Somatic embryogenesis	*AcSERK1*	Induced	*SERK* activation links auxin signaling to embryogenic competence	Ma et al. (2012)
	Ocotea catharinensis	Embryogenic cells	*OcWUS, OcWOX4*	Restricted expression	Spatially constrained auxin-responsive *WOX* module	Santa-Catarina et al. (2012)
	Gossypium hirsutum	Callus	*GhAGL15*	Induced	Auxin-mediated activation of maturation-associated regulators	Yang et al. (2014)
	Saccharum officinarum	Somatic embryogenesis, embryogenic callus, shoot formation	*SoBBM, SoLEC, SoWUS*	Upregulated	Coordinated auxin-mediated network activation across embryogenic regulators	Maulidiya et al. (2020)
	Musa acuminata	Embryogenic callus	*MaBBM1/2, MaWUS2*	Upregulated	Auxin-induced activation of embryogenic hub genes promotes somatic embryogenesis competence	Shivani et al. (2017)
	Vitis vinifera	Somatic embryogenesis	*VvSERK1, VvL1L*	Downregulated (stage-specific reactivation)	Context-dependent auxin rewiring between primary and secondary embryogenesis	Maillot et al. (2009)
TDZ	*Brassica juncea*	Callus	*BjuAux/IAA, BjuERF1/2/6*	Upregulated	Cytokinin-ethylene interaction module activated	Lu et al. (2020)

(continued)

Table 6.1 (continued)

Growth hormones	Species	Developmental context	Target developmental regulators	Transcriptional response	Network-level interpretation	References
IAA	*Coffea canephora*	Somatic embryogenesis	*CcYUC, CcTAA1*	Upregulated	Positive feedback on auxin biosynthesis module	Ayil-Gutiérrez et al. (2013)
NAA	*Arabidopsis thaliana*	Shoot regeneration	*AtLEC2*	Overexpressed	Auxin-cytokinin combinatorial regulation of *LEC* module	Li et al. (2019)
BAP	*Arabidopsis thaliana*	Shoot regeneration	*AtLEC2*	Overexpressed	Cytokinin-mediated shift toward embryogenic identity	Li et al. (2019)
	Lilium lancifolium	Bulbil development	*LlWOX9, LlWOX11*	Strong induction	Cytokinin-driven temporal *WOX* activation and meristem specification	He et al. (2022)
ABA	*Triticum aestivum*	Somatic embryogenesis	*TaSERK1/2*	Induced	Stress hormone coupling enhances embryogenic signaling	Singla et al. (2008)
GA_3	*Medicago truncatula*	Callus and somatic embryogenesis	*MtBBM1*	Downregulated	Gibberellin antagonism of embryogenic network	Igielski and Kępczyńska (2017)
		Embryogenesis	*MtCPS*, *MtGA$_3$OX1/2*	Upregulated	GA biosynthesis genes activated during embryo development	
2,4-D + BAP	*Zea mays*	Embryogenesis	*ZmSERK1/2* (up), *ZmSERK3* (down)	Differential regulation	Combinatorial hormone-driven subfunctionalization within *SERK* family	Zhang et al. (2011)
	Passiflora edulis	Somatic embryos progression	*PeSERK1*, *PeSERK2*, *PeSERK3*	Developmental stage-dependent	Sequential activation reflecting temporal gene regulatory network restructuring	Rocha et al. (2016)

BAP + NAA	*Melocactus glaucescens*	Shoot regeneration	*MgSERK1*	Induced	Synergistic hormone-dependent *SERK* activation	Torres-Silva et al. (2021)
	Gossypium hirsutum	Callus	*GhAGL15–1/15–3*	Induced	Cytokinin-auxin interaction influencing maturation network	Yang et al. (2014)
Auxin + cytokinin	*Triticum aestivum*	Root formation from callus	*TaWOX5*	Induced	Stem cell niche specification via combinatorial signaling	Zhao et al. (2014)
NAA + BAP	*M. truncatula*	Somatic embryogenesis and root	*MtSERK1*, *MtWOX11*-like, *MtPIN10*	Induced	Hormonal coordination of auxin transport and stem cell regulators	Nolan et al. (2003); Tvorogova et al. (2015)
Paclobutrazol (GA inhibitor)	*M. truncatula*	Callus and somatic embryogenesis	*MtBBM1*	Upregulated	Relief of GA repression restores embryogenic competence	Igielski and Kępczyńska (2017)
NAA + BAP + IAA	*M. truncatula*	Callus, somatic embryogenesis, meristem formation	*MtWUS*, *MtWOX5*	Induced	Integrated auxin-cytokinin module specifying stem cell identity	Chen et al. (2009)

2,4-D 2,4-dichlorophenoxyacetic acid, *ABA* abscisic acid, *AGL15* AGAMOUS-Like 15, *AHL* Hook Motif Nuclear-Localized, *BAP* 6-Benzylaminopurine, *BBM* BABY BOOM, *CPS* Copalyl Diphosphate Synthase, *ERF1* Ethylene Response Factor 1, *GA* gibberellic acid, *GA_3OX1* gibberellin 3-beta-dioxygenase 1, *IAA* indole-3-acetic acid, *LEC* LEAFY COTYLEDON, *L1L* LEC1-LIKE, *NAA* naphthalene acetic acid, *PIN* PIN-FORMED, *SERK* SOMATIC EMBRYOGENESIS RECEPTOR-LIKE KINASE, *TAA1* Tryptophan Aminotransferase of *Arabidopsis* 1, TDZ thidiazuron, *YUC* YUCCA, *WOX* WUSCHEL-RELATED HOMEOBOX, *WIND* WOUND-INDUCED DEDIFFERENTIATION, *WUS*- WUSCHEL cascades, thereby determining whether orthologous genes promote or destabilize embryogenic fate. Similarly, in *Passiflora edulis*, members of the *SERK* family exhibit developmentally regulated expression patterns. *PeSERK1* is not expressed in explants prior to induction; *PeSERK2* shows limited expression during early somatic embryogenesis and *PeSERK3* becomes fully expressed at the globular stage (Rocha et al. 2016). This sequential activation pattern underscores a hormone-sensitive developmental switch rather than constitutive gene activity. Indeed, auxin-dependent *SERK* expression has also been documented in carrot and *Arabidopsis*, where *SERK* overexpression stimulates somatic embryogenesis competence highlighting auxin, signaling reorganizes receptor kinase-mediated embryogenic pathways (Hecht et al. 2001). Woody perennials further exemplify gene regulatory network plasticity under hormonal influence. In *Populus tomentosa*, overexpression of *PtoWOX5a* increases root formation but reduces leaf number, revealing pleiotropic network effects and altered organogenic balance (Liu et al. 2014). In coffee (*Coffea canephora*), auxin-induced upregulation of *CcYUC* (YUCCA) and *CcTAA1* (Tryptophan Aminotransferase of *Arabidopsis* (1) enhances endogenous IAA biosynthesis during somatic embryogenesis while *CcLEC1*, *CcBBM1* and *CcWOX4* exhibit stage-specific transcriptional activation during globular and heart stages (Ayil-Gutiérrez et al. 2013). These patterns indicated that endogenous auxin accumulation reshapes downstream transcriptional hierarchies.

The role of plant growth hormones in gene regulatory network rewiring becomes particularly evident when comparing species responses to identical hormonal treatments. The synthetic auxin 2,4-D (2,4-dichlorophenoxyacetic acid) activates a group of embryogenesis-related transcription factors including *WUS*, *LEC*, L1 Layer-Specific Homeobox (*L1L*), *BBM*, *SERK*, *and AGL15* across multiple species (Winnicki 2020). In banana (*Musa acuminata*) and sugarcane (*Saccharum officinarum*), 2,4-D induces *BBM, WUS* and *LEC* orthologs to promote embryogenic cell suspension and somatic embryogenesis formation (Shivani et al. 2017; Maulidiya et al. 2020). Conversely, in grapevine (*Vitis vinifera*), 2,4-D downregulates *VvSERK1* and *VvL1L* during primary somatic embryogenesis but upregulates them during secondary embryogenesis, demonstrating context-dependent expression shifts (Maillot et al. 2009). Cytokinin similarly drives network reconfiguration. In *Medicago truncatula*, early *WUS* expression is cytokinin-dependent, with maximal transcription occurring one week after cytokinin application during stem cell induction (Rose 2019). In *Lilium lancifolium*, benzyl adenine (BA) treatment significantly elevates *LlWOX9* and *LlWOX11* expression during bulbil formation, accompanied by modulation of cytokinin biosynthesis genes i.e., isopentenyl transferase 1 / isopentenyl transferase 5 (*IPT1/5*), activation genes i.e., *Lonely Guy 1 / 3 / 5 / 7* (*LOG1/3/5/7*), and degradation genes i.e., *Cytokinin Oxidase/Dehydrogenase 4* (*CKX4*) (He et al. 2022). These coordinated changes in hormone metabolism genes revealed a multilayered feedback loop that reprograms developmental trajectories.

Auxin-cytokinin crosstalk further exemplifies gene regulatory network plasticity. In *Arabidopsis*, auxin-induced callus formation involves differential expression of *Auxin Response Factor 5* (*ARF5*) and *Ethylene Response DNA-Binding Factor 3* (*EDF3*), genes also upregulated in wounding-induced callus, suggesting convergence between hormonal and stress-responsive networks (Iwase et al. 2011). Moreover, *WOX* family members interact with cytokinin response regulators and auxin biosynthetic genes such as *TAA1*, disrupting negative feedback loops and promoting organogenic competence (Lee et al. 2022). In hybrid sweetgum (*Liquidambar styraciflua* × *L. formosana*), auxin-responsive genes (*WOX9, PIN1-like, AUX1, ARF17, ARF18*) are upregulated in embryogenic callus but downregulated in somatic embryos, illustrating stage-dependent hormonal rewiring (Qi et al. 2021). These findings demonstrated that plant growth hormones act as dynamic modulators of expression divergence. While orthologous developmental regulatory genes are broadly conserved, species-specific promoter architectures, chromatin landscapes, and hormone sensitivities determine how these genes are integrated into developmental networks. Auxin, cytokinin, and gibberellin signals alter transcription factor hierarchies, receptor kinase activation, and downstream morphogenetic pathways, thereby rewiring gene regulatory networks in a lineage-specific manner. Consequently, differences in regenerative ability and totipotent competence among plant species are mainly due to evolutionary changes in hormone-responsive regulatory networks, rather than simply differences in the types or number of genes they possess.

6.4 Evolutionary Basis of Regenerative Capacity

The cross-species conservation of *BBM*, *LEC*, *WUS/WOX*, and *SERK* orthologs indicates that regenerative competence evolved early in seed plant evolution as part of embryogenesis control mechanisms. Their recurrent recruitment in vitro likely reflects co-option of ancestral embryonic programs. Gymnosperm evidence (e.g., *Picea*, *Pinus*, *Larix*) demonstrates that *WOX* and *BBM*-like genes regulate embryogenic mass formation and early embryo patterning (Rupps et al. 2016; Zhu et al. 2016). This supports the hypothesis that totipotency regulators originated in the control of zygotic embryogenesis and were later redistributed in somatic contexts. The widespread ability of ectopic *BBM* or *WUS* expression to induce somatic embryogenesis in diverse angiosperms including maize, cotton, tobacco, and *Arabidopsis* (Boutilier 2007; Srinivasan et al. 2007) suggests that developmental plasticity is an ancestral trait retained across lineages but differentially modulated.

Phylogenetic comparison reveals patterns of regeneration efficiency aligned with lineage-specific regulatory tuning. Dicots generally display strong embryogenic responses upon developmental regulatory gene overexpression (e.g., *Arabidopsis*, *Nicotiana*, *Gossypium*, *Brassica*) (Belide et al. 2013; Guo et al. 2013). Monocots such as maize and rice require combined hormonal and genetic modulation. Woody perennials (e.g., *Populus*, *Citrus*, *Coffee*) have shown the stage-dependent gene activation and often lower baseline regeneration capacity, requiring precise temporal regulation of ortholog expression (Liu et al. 2018). Meanwhile, gymnosperms exhibit embryogenic gene conservation but prolonged and tightly staged expression patterns, reflecting evolutionary constraints associated with long life cycles and complex embryo development. The comparative evidence demonstrates that plant totipotency is governed by a conserved core module of developmental regulators *LEC*, *BBM*, *WUS/WOX*, *SERK*, and *AGL* families whose orthologs function across angiosperms and gymnosperms. However, evolutionary divergence in expression dynamics, network connectivity, and hormonal integration has generated species-specific regeneration phenotypes. Thus, regenerative competence reflects both evolutionary conservation and lineage-specific regulatory innovation, providing a framework for understanding recalcitrance and guiding synthetic enhancement of regeneration systems in crop biotechnology.

6.5 Developmental Regulatory Genes in Recalcitrant Species

Forest trees, particularly long-lived woody species characterized by complex and highly heterozygous genomes, present substantial challenges for somatic regeneration and genetic transformation. These limitations are collectively described as recalcitrance and are typically manifested as low regeneration efficiency and strong genotype dependence (Guo et al. 2020; Long et al. 2022).

6.5.1 Low Regeneration Efficiency and Genotype Dependence

One of the most prominent challenges in forest tree biotechnology is the pronounced variation in regenerative capacity among different genotypes. Even under identical in vitro culture conditions, some genotypes display high embryogenic competence, whereas others fail to regenerate entirely. For example, comparative transcriptomic studies in *Gossypium hirsutum* have revealed extensive differences between highly embryogenic and recalcitrant genotypes. These differences are largely associated with variations in transcription factor activity and alternative splicing patterns that influence embryogenic potential (Guo et al. 2020). The recalcitrant nature of many forest species is influenced by multiple interacting factors, including species-specific biological characteristics, genetic variability among individuals, the composition and balance of plant growth hormones, and the physical and chemical properties of the in vitro culture environment. The complex interplay of these variables reflects the intricate cellular and molecular networks that govern the acquisition of totipotency in plant cells (Long et al. 2022; Chen et al. 2024). Despite decades of research, the molecular mechanisms underlying dedifferentiation, callus formation, and the acquisition of totipotency remain only partially understood. This knowledge gap complicates the development of universally applicable regeneration protocols. The challenge is particularly pronounced in forest trees due to their extended life cycles, high genetic heterogeneity, and comparatively limited regenerative plasticity (Long et al. 2022).

6.5.2 *Molecular and Physiological Constraints*

Emerging evidence indicates that several molecular and physiological disruptions contribute to regeneration recalcitrance in woody species. These include imbalances in endogenous hormone homeostasis, epigenetic modifications such as DNA methylation and histone modifications, and dysregulated expression of developmental regulatory genes (Fehér 2019). In recent years, the application of small-molecule regulators, including epigenetic inhibitors and antioxidants, has demonstrated potential for enhancing somatic embryogenesis efficiency. Nevertheless, translating these experimental findings into scalable and reliable strategies for forest tree improvement remains a significant challenge.

6.6 Synthetic Approaches to Overcome Species-Specific Constraints Through Developmental Regulatory Genes

The barriers associated with somatic regeneration and genetic transformation in recalcitrant forest species may be overcome through targeted manipulation of developmental regulatory genes. Conventional regeneration methods are frequently hindered by low callus induction rates, extensive tissue browning, and poor somatic embryo maturation and germination. Elevated levels of phenolic compounds and disruptions in endogenous hormonal signaling pathways, particularly those involving auxins and cytokinins, are key contributors to regeneration constraints in woody plants. For example, somatic embryogenesis in *Quercus* subspecies exhibits strong genotype dependence, which is associated with transcriptional repression of essential embryogenic regulators and the accumulation of stress-related secondary metabolites. Additionally, robust epigenetic barriers, including dynamic patterns of DNA methylation and histone modification, are closely linked with developmental recalcitrance (Wang et al. 2025). Synthetic biology offers a transformative paradigm for enhancing somatic regeneration and genetic transformation in recalcitrant spp. by engineering modular, controllable gene circuits. This approach aims to achieve finely tuned control over regeneration processes, moving beyond the limitations of single-gene strategies.

6.6.1 *Enhancing Embryogenic Competence Through Developmental Regulatory Genes*

Developmental regulatory genes such as *WUS*, *BBM*, and the Growth-Regulating Factor-Interacting Factor (*GRF–GIF*) transcriptional complex have emerged as powerful molecular tools for promoting embryogenic competence and stimulating cellular proliferation, even in highly recalcitrant tissues. For example, overexpression of *HbGRF4* or a chimeric *HbGRF4–GIF1* construct significantly increased somatic embryo formation in *Hevea brasiliensis*, effectively overcoming genotype-related limitations by activating key transcriptional networks involved in embryogenic development. Similarly, the overexpression of *Larix kaempferi Cyclin-Dependent Kinase B1;2* (*LaCDKB1;2*) has been shown to improve the yield and quality of somatic embryos in *Larix kaempferi* through its regulatory role in cell cycle progression (Capote et al. 2019; Kong et al. 2020).

6.6.2 Functional Validation and Strategic Advances

Functional characterization of developmental regulatory genes in recalcitrant forest tree species remains challenging due to the inherent difficulties associated with regeneration and transformation. To address these limitations, several methodological approaches have been developed to facilitate gene function analysis.

6.6.2.1 Transient Expression Approaches

Transient expression systems, including agroinfiltration, particle bombardment, and protoplast-based assays, provide rapid and efficient platforms for evaluating the activity of candidate developmental genes without requiring stable transformation. The use of reporter genes such as Green Fluorescent Protein (*GFP*) and β-Glucuronidase (*GUS*) enables the visualization of spatial and temporal patterns of regulatory gene expression during callus induction and somatic embryogenesis. These systems offer valuable insights into the early molecular events involved in developmental reprogramming (Li et al. 2025).

6.6.2.2 Stable Transformation Systems

Although technically demanding, stable genetic transformation remains the most reliable method for comprehensive functional characterization of regulatory genes. In several forest species, including *Pinus pinaster* and *Quercus suber*, *Agrobacterium*-mediated transformation coupled with somatic embryogenesis-based regeneration systems has provided an effective platform for investigating gene function and regulatory networks involved in embryogenic development (Wang et al. 2025).

6.6.2.3 Genome Editing Approaches

The advent of clustered regularly interspaced short palindromic repeats/CRISPR-associated protein 9 (CRISPR/Cas9) genome editing technology has introduced new opportunities for precise genetic manipulation in recalcitrant species. This approach enables the generation of targeted gene knockouts, allelic modifications, or controlled gene overexpression lines. Genome editing has already been applied in species such as larch and other conifers to investigate the functional roles of regulatory genes in regeneration processes. When combined with transcriptomic, epigenomic, and protein-protein interaction analyses, CRISPR-based approaches allow a more comprehensive dissection of the molecular networks governing regeneration competence (Li et al. 2025). Furthermore, developmental regulatory genes can serve as co-transformation or preparatory tools to enhance regenerative capacity following transformation, thereby addressing one of the most significant bottlenecks in the genetic improvement of recalcitrant spp.

6.7 Conclusion

Developmental regulators such as *LEC*, *BBM*, *WUS*, *WOX*, and *AGL* genes represent evolutionarily conserved components of plant regeneration. Despite this conservation, substantial variation exists in regenerative capacity among species, largely driven by evolutionary divergence in regulatory DNA sequences, chromatin organization, hormone responsiveness, and gene regulatory network architecture. These regulatory differences underlie the widespread phenomenon of regeneration recalcitrance observed in many economically important crops and recalcitrant species. Advances in plant biotechnology are increasingly enabling the targeted manipulation of developmental regulatory pathways to overcome these limitations. Synthetic approaches that involve the controlled expression of key developmental regulators, together with modern genome editing technologies such as CRISPR/Cas systems, provide powerful tools for enhancing regeneration competence in otherwise recalcitrant species. The integration of evolutionary developmental insights with synthetic biology strategies therefore offers a promising framework for the rational design of improved regeneration systems.

References

Ayil-Gutiérrez B, Galaz-Ávalos RM, Peña-Cabrera E, Loyola-Vargas VM (2013) Dynamics of the concentration of IAA and some of its conjugates during the induction of somatic embryogenesis in Coffea canephora. Plant Signal Behav 8(11):e26998

Belide S, Zhou XR, Kennedy Y, Lester G, Shrestha P, Petrie JR, Singh SP (2013) Rapid expression and validation of seed-specific constructs in transgenic LEC2 induced somatic embryos of Brassica napus. Plant Cell Tissue Organ Cult 113(3):543–553

Boutilier K (2007) Heterologous expression of the BABY BOOM AP2/ERF transcription factor enhances the regeneration capacity of tobacco (Nicotiana tabacum L.). Planta 225(2):341–351

Capote T, Usié A, Barbosa P, Ramos M, Morais-Cecílio L, Gonçalves S (2019) Transcriptome dynamics of cork oak (Quercus suber) somatic embryogenesis reveals active gene players in transcription regulation and phytohormone homeostasis of embryo development. Tree Genet Genomes 15(4):52

Chen SK, Kurdyukov S, Kereszt A, Wang XD, Gresshoff PM, Rose RJ (2009) The association of homeobox gene expression with stem cell formation and morphogenesis in cultured Medicago truncatula. Planta 230(4):827–840

Chen C, Hu Y, Ikeuchi M, Jiao Y, Prasad K, Su YH, Xiao J, Xu L, Yang W, Zhao Z, Wang JW (2024) Plant regeneration in the new era: from molecular mechanisms to biotechnology applications. Sci China Life Sci 67(7):1338–1367

Fambrini M, Usai G, Pugliesi C (2022) Induction of somatic embryogenesis in plants: different players and focus on WUSCHEL and WUS-RELATED HOMEOBOX (WOX) transcription factors. Int J Mol Sci 23(24):15950

Fehér A (2019) Callus, dedifferentiation, totipotency, somatic embryogenesis: what these terms mean in the era of molecular plant biology? Front Plant Sci 10:536

Guo F, Liu C, Xia H, Bi Y, Zhao C, Zhao S, Hou L, Li F, Wang X (2013) Induced expression of AtLEC1 and AtLEC2 differentially promotes somatic embryogenesis in transgenic tobacco plants. PLoS One 8(8):e71714

Guo H, Guo H, Zhang L, Fan Y, Wu J, Tang Z, Zhang Y, Fan Y, Zeng F (2020) Dynamic transcriptome analysis reveals uncharacterized complex regulatory pathway underlying genotype-recalcitrant somatic embryogenesis transdifferentiation in cotton. Genes 11(5):519

He G, Cao Y, Wang J, Song M, Bi M, Tang Y, Xu L, Ming J, Yang P (2022) WUSCHEL-related homeobox genes cooperate with cytokinin to promote bulbil formation in Lilium lancifolium. Plant Physiol 190(1):387–402

Hecht V, Vielle-Calzada JP, Hartog MV, Schmidt ED, Boutilier K, Grossniklaus U, de Vries SC (2001) The Arabidopsis SOMATIC EMBRYOGENESIS RECEPTOR KINASE 1 gene is expressed in developing ovules and embryos and enhances embryogenic competence in culture. Plant Physiol 127(3):803–816

Hedman H, Zhu T, Von Arnold S, Sohlberg JJ (2013) Analysis of the WUSCHEL-RELATED HOMEOBOX gene family in the conifer Picea abies reveals extensive conservation as well as dynamic patterns. BMC Plant Biol 13(1):89

Horstman A, Li M, Heidmann I, Weemen M, Chen B, Muino JM, Angenent GC, Boutilier K (2017) The BABY BOOM transcription factor activates the LEC1-ABI3-FUS3-LEC2 network to induce somatic embryogenesis. Plant Physiol 175(2):848–857

Hu H, Xiong L, Yang Y (2005) Rice SERK1 gene positively regulates somatic embryogenesis of cultured cell and host defense response against fungal infection. Planta 222(1):107–117

Hu B, Zhang G, Liu W, Shi J, Wang H, Qi M, Li J, Qin P, Ruan Y, Huang H, Xu L (2017) Divergent regeneration-competent cells adopt a common mechanism for callus initiation in angiosperms. Regeneration 4(3):132–139

Igielski R, Kępczyńska E (2017) Gene expression and metabolite profiling of gibberellin biosynthesis during induction of somatic embryogenesis in Medicago truncatula Gaertn. PLoS One 12(7):e0182055

Iwase A, Mitsuda N, Koyama T, Hiratsu K, Kojima M, Arai T, Inoue Y, Seki M, Sakakibara H, Sugimoto K, Ohme-Takagi M (2011) The AP2/ERF transcription factor WIND1 controls cell dedifferentiation in Arabidopsis. Curr Biol 21(6):508–514

Jiang Y, Wei X, Zhu M, Zhang X, Jiang Q, Wang Z, Cao Y, An X, Wan X (2024) Developmental regulators in promoting genetic transformation efficiency in maize and other plants. Curr Plant Biol 40:100383

Karami O, Rahimi A, Mak P, Horstman A, Boutilier K, Compier M, van der Zaal B, Offringa R (2021) An *Arabidopsis* AT-hook motif nuclear protein mediates somatic embryogenesis and coinciding genome duplication. Nat Commun 12(1):2508

Kong J, Martin-Ortigosa S, Finer J, Orchard N, Gunadi A, Batts LA, Thakare D, Rush B, Schmitz O, Stuiver M, Pacheco-Villalobos D (2020) Overexpression of the transcription factor GROWTH-REGULATING FACTOR 5 improves transformation of dicot and monocot species. Front Plant Sci 11:572319

Ledwoń A, Gaj MD (2009) LEAFY COTYLEDON2 gene expression and auxin treatment in relation to embryogenic capacity of Arabidopsis somatic cells. Plant Cell Rep 28(11):1677–1688

Lee K, Kim JH, Park OS, Jung YJ, Seo PJ (2022) Ectopic expression of WOX5 promotes cytokinin signaling and de novo shoot regeneration. Plant Cell Rep 41(12):2415–2422

Li K, Wang J, Liu C, Li C, Qiu J, Zhao C et al (2019) Expression of AtLEC2 and AtIPTs promotes embryogenic callus formation and shoot regeneration in tobacco. BMC Plant Biol 19(1):314

Li Y, Ma W, Qi S, Ma H, Li H, Bao F, Fan Y, Zhan D, Pang Z, Zhao J, Zhang J (2025) Genome-wide characterization of the WRKY gene family and the role of LsfWRKY29 in regulating somatic embryogenesis in hybrid sweetgum (Liquidambar styraciflua× L. formosana). Int J Biol Macromol 293:139287

Liu B, Wang L, Zhang J, Li J, Zheng H, Chen J, Lu M (2014) WUSCHEL-related Homeobox genes in Populus tomentosa: diversified expression patterns and a functional similarity in adventitious root formation. BMC Genomics 15(1):296

Liu Z, Ge XX, Qiu WM, Long JM, Jia HH, Yang W, Dutt M, Wu XM, Guo WW (2018) Overexpression of the CsFUS3 gene encoding a B3 transcription factor promotes somatic embryogenesis in citrus. Plant Sci 277:121–131

Long Y, Yang Y, Pan G, Shen Y (2022) New insights into tissue culture plant-regeneration mechanisms. Front Plant Sci 13:926752

Lu H, Xu P, Hu K, Xiao Q, Wen J, Yi B, Ma C, Tu J, Fu T, Shen J (2020) Transcriptome profiling reveals cytokinin promoted callus regeneration in Brassica juncea. Plant Cell Tissue Organ Cult 141(1):191–206

Ma J, He Y, Wu C, Liu H, Hu Z, Sun G (2012) Cloning and molecular characterization of a SERK gene transcriptionally induced during somatic embryogenesis in Ananas comosu s cv. Shenwan. *Plant Molecular Biology Reporter* 30(1):195–203

Maillot P, Lebel S, Schellenbaum P, Jacques A, Walter B (2009) Differential regulation of SERK, LEC1-like and pathogenesis-related genes during indirect secondary somatic embryogenesis in grapevine. Plant Physiol Biochem 47(8):743–752

Marand AP, Eveland AL, Kaufmann K, Springer NM (2023) Cis-regulatory elements in plant development, adaptation, and evolution. Annu Rev Plant Biol 74(1):111–137

Maulidiya AUK, Sugiharto B, Dewanti P, Handoyo T (2020) Expression of somatic embryogenesis-related genes in sugarcane (Saccharum officinarum L.). J Crop Sci Biotechnol 23(3):207–214

Mookkan M, Nelson-Vasilchik K, Hague J, Zhang ZJ, Kausch AP (2017) Selectable marker independent transformation of recalcitrant maize inbred B73 and sorghum P898012 mediated by morphogenic regulators BABY BOOM and WUSCHEL2. Plant Cell Rep 36(9):1477–1491

Nolan KE, Irwanto RR, Rose RJ (2003) Auxin up-regulates MtSERK1 expression in both Medicago truncatula root-forming and embryogenic cultures. Plant Physiol 133(1):218–230

Orłowska A, Kępczyńska E (2018) Identification of Polycomb repressive Complex1, Trithorax group genes and their simultaneous expression with WUSCHEL, WUSCHEL-related Homeobox5 and SHOOT MERISTEMLESS during the induction phase of somatic embryogenesis in Medicago truncatula Gaertn. Plant Cell Tissue Organ Cult 134(3):345–356

Orłowska A, Igielski R, Łagowska K, Kępczyńska E (2017) Identification of LEC1, L1L and Polycomb repressive complex 2 genes and their expression during the induction phase of Medicago truncatula Gaertn. Somatic embryogenesis. Plant Cell Tissue Organ Cult 129(1):119–132

Qi S, Zhao R, Yan J, Fan Y, Huang C, Li H, Chen S, Zhang T, Kong L, Zhao J, Zhang J (2021) Global transcriptome and coexpression network analyses reveal new insights into somatic embryogenesis in hybrid sweetgum (Liquidambar styraciflua× Liquidambar formosana). Front Plant Sci 12:751866

Rocha DI, Pinto DLP, Vieira LM, Tanaka FAO, Dornelas MC, Otoni WC (2016) Cellular and molecular changes associated with competence acquisition during passion fruit somatic embryogenesis: ultrastructural characterization and analysis of SERK gene expression. Protoplasma 253(2):595–609

Rose RJ (2019) Somatic embryogenesis in the Medicago truncatula model: cellular and molecular mechanisms. Front Plant Sci 10:267

Rupps A, Raschke J, Rümmler M, Linke B, Zoglauer K (2016) Identification of putative homologs of Larix decidua to Babyboom (BBM), leafy cotyledon1 (LEC1), Wuschel-related Homeobox2 (WOX2) and somatic embryogenesis receptor-like kinase (SERK) during somatic embryogenesis. Planta 243(2):473–488
Santa-Catarina C, de Oliveira RR, Cutri L, Floh EI, Dornelas MC (2012) WUSCHEL-related genes are expressed during somatic embryogenesis of the basal angiosperm Ocotea catharinensis Mez.(Lauraceae). Trees 26(2):493–501
Shivani, Awasthi P, Sharma V, Kaur N, Kaur N, Pandey P, Tiwari S (2017) Genome-wide analysis of transcription factors during somatic embryogenesis in banana (Musa spp.) cv. Grand Naine. PLoS One 12(8):e0182242
Singla B, Khurana JP, Khurana P (2008) Characterization of three somatic embryogenesis receptor kinase genes from wheat, Triticum aestivum. *Plant Cell Reports* 27(5):833–843
Srinivasan C, Liu Z, Heidmann I, Supena EDJ, Fukuoka H, Joosen R, Srinivasan C, Liu Z, Heidmann I, Supena EDJ, Fukuoka H, Joosen R et al (2007) Heterologous expression of the BABY BOOM AP2/ERF transcription factor enhances the regeneration capacity of tobacco (Nicotiana tabacum L.). Planta 225(2):341–351
Torres-Silva G, Correia LNDF, Koehler AD, Batista DS, Faria DV, Resende SV, Strickler SR, Fouracre J, Romanel E, Specht CD, Otoni WC (2021) Expression of Melocactus glaucescens SERK1 sheds new light on the mechanism of areolar activation in cacti. Plant Cell Tissue Organ Cult 147(3):437–451
Tvorogova VE, Lebedeva MA, Lutova LA (2015) Expression of WOX and PIN genes during somatic and zygotic embryogenesis in Medicago truncatula. Russ J Genet 51(12):1189–1198
Wang Y, Wang JH, Guo PR, Peng J, Xie CZ, Shi YD, Wu YH, Liao DZ, Sheng S (2025) Developmental regulatory genes in recalcitrant forest trees: advances in somatic regeneration and genetic transformation. Front Plant Sci 16:1689705
Winnicki K (2020) The winner takes it all: auxin—the main player during plant embryogenesis. Cells 9(3):606
Xiao J, Jin R, Wagner D (2017) Developmental transitions: integrating environmental cues with hormonal signaling in the chromatin landscape in plants. Genome Biol 18(1):88
Yang Z, Li C, Wang Y, Zhang C, Wu Z, Zhang X, Liu C, Li F (2014) GhAGL15s, preferentially expressed during somatic embryogenesis, promote embryogenic callus formation in cotton (Gossypium hirsutum L.). Mol Gen Genomics 289(5):873–883
Yu Y, Yu H, Peng J, Yao WJ, Wang YP, Zhang FL, Wang SR, Zhao Y, Zhao XY, Zhang XS, Su YH (2024) Enhancing wheat regeneration and genetic transformation through overexpression of TaLAX1. Plant Comm 5(5):100738
Zhang S, Liu X, Lin Y, Xie G, Fu F, Liu H, Wang J, Gao S, Lan H, Rong T (2011) Characterization of a ZmSERK gene and its relationship to somatic embryogenesis in a maize culture. Plant Cell Tissue Organ Cult 105(1):29–37
Zhao S, Jiang QT, Ma J, Zhang XW, Zhao QZ, Wang XY, Wang CS, Cao X, Lu ZX, Zheng YL, Wei YM (2014) Characterization and expression analysis of WOX5 genes from wheat and its relatives. Gene 537(1):63–69
Zheng W, Zhang X, Yang Z, Wu J, Li F, Duan L, Liu C, Lu L, Zhang C, Li F (2014) AtWuschel promotes formation of the embryogenic callus in Gossypium hirsutum. PLoS One 9(1):e87502
Zhu SP, Wang J, Ye JL, Zhu AD, Guo WW, Deng XX (2014) Isolation and characterization of LEAFY COTYLEDON 1-LIKE gene related to embryogenic competence in Citrus sinensis. Plant Cell Tissue Organ Cult 119(1):1–13
Zhu T, Moschou PN, Alvarez JM, Sohlberg JJ, von Arnold S (2016) WUSCHEL-RELATED HOMEOBOX 2 is important for protoderm and suspensor development in the gymnosperm Norway spruce. BMC Plant Biol 16(1):19

Chapter 7
Synthetic Modulation and Future Directions in Regeneration Engineering

Abstract Advances in plant synthetic biology are transforming regeneration from a largely empirical, hormone-driven process into a programmable developmental system. Recent technologies such as clustered regularly interspaced short palindromic repeats-based transcriptional activation (CRISPRa), synthetic promoters, orthogonal regulatory systems, artificial gene circuits, and epigenetic editing technologies allow precise control over the activation of morphogenic regulators involved in plant regeneration. These innovations allow the modulation of gene expression spatially, temporally, and quantitatively, thereby improving the efficiency and predictability of somatic embryogenesis and organogenesis. Synthetic gene circuits that include toggle switches, Boolean logic gates, pulse generators, and feedback loops provide a framework for controlling cell fate transitions and stabilizing totipotent states. Furthermore, targeted epigenetic reprogramming using CRISPR-based epigenetic editors can modify chromatin states at developmental loci, overcoming epigenetic barriers that often limit regeneration. The integration of these technologies may lead to programmable regeneration platforms capable of enabling genotype-independent transformation, *in planta* regeneration, and rapid crop improvement when combined with accelerated breeding strategies. However, the introduction of such systems also raises ethical, regulatory, and commercial considerations related to biosafety, intellectual property, and genome editing policies. Regeneration engineering is offering new opportunities to enhance transformation efficiency, expand the range of transformable crops, and accelerate the development of improved agricultural varieties.

Keywords CRISPR · Crop improvement · Gene circuit · Plant regeneration engineering · Synthetic biology · Synthetic promoters

P. Kumar, *Totipotency in Plants*, SpringerBriefs in Plant Science,
https://doi.org/10.1007/978-981-92-1631-4_7

7.1 Introduction

Plant regeneration has been one of the most powerful but unpredictable processes in plant biotechnology. Traditionally, regeneration depended on specifically regulating plant hormones in tissue culture medium to encourage differentiated cells to generate callus and eventually new shoots or embryos. While this method has been successful in model species such as *Arabidopsis thaliana*, it is frequently inefficient and highly genotype-dependent in major crops such as *Oryza sativa* and *Zea mays*. As a result, regeneration remains a major barrier in plant transformation and genome editing (Debernardi et al. 2020). However, recent developments in synthetic biology and molecular biology have opened a new chapter in the understanding of regeneration, which is now increasingly seen as a genetic process that can be programmed rather than just a hormone-driven reaction. Master regulatory genes govern cell fate transitions, which are crucial to regeneration. Transcription factors such as *WUSCHEL* (*WUS*), *BABY BOOM* (*BBM*), and *LEAFY COTYLEDON 1/LEAFY COTYLEDON 2* (*LEC1/LEC2*) act as developmental regulators. When activated, they can redirect somatic cells toward embryogenic or meristematic states. Overexpression of *BBM* or *WUS* significantly enhances regeneration efficiency, particularly in recalcitrant crops (Lowe et al. 2016). These findings highlighted that instead of indirectly stimulating regeneration through hormones, directly activating key regulatory genes can be significantly more efficient. One of the most promising techniques for facilitating this change is CRISPRa. Unlike conventional CRISPR systems that edit deoxyribonucleic acid (DNA) sequences, CRISPRa uses a catalytically inactive Cas protein (dCas9) fused to transcriptional activation domains. The complex, guided by specific ribonucleic acid (RNA) molecules, binds to promoter regions and increases expression of target genes without altering the genome. This technique allows precise, adjustable, and reversible activation of endogenous developmental regulators (Lowder et al. 2018). CRISPRa has numerous benefits for regeneration engineering, including the ability to minimize genomic disruption, allows multiplex activation of several genes simultaneously, and can be placed under inducible control. For example, temporary activation of *WUS* and *BBM* through inducible systems can stimulate regeneration while avoiding long-term developmental abnormalities such as ectopic embryo formation. Beyond transcriptional activation, synthetic promoter engineering provides an additional level of refinement. Synthetic promoters can be designed to activate gene expression in specific tissues or in response to defined stimuli such as chemicals, heat, or light. This precision is crucial because uncontrolled expression of morphogenic regulators can lead to sterility or abnormal growth. Synthetic promoters enhance both efficiency and biosafety by regulating expression in space and time (Liu and Stewart 2016). Orthogonal regulatory systems, designed to function independently of the plant's native transcriptional machinery, further improve predictability by reducing random interactions with endogenous gene networks.

One of the most recent advancements in regenerative engineering is the development of artificial gene circuits. These include the principles from engineering and computer logic, allowing scientists to regulate when and how regeneration occurs. Instead of relying on unpredictable hormone responses, regeneration can be programmed to occur only under specific conditions. For example, an AND gate design could trigger embryogenesis only when both wound signal and an external inducer are present. Toggle switches may retain cells in an embryogenic state when necessary, and feedback mechanisms guarantee that the process ends at the appropriate moment. Overall, circuit-based techniques improve regeneration control, reliability, and predictability (Koukara and Papadopoulou 2023; Khan and Lister 2025). Epigenetic reprogramming is another important regulatory mechanism in regenerative engineering. To recover embryogenic potential, a cell chromatin structure must be modified. In differentiated cells, repressive markers like trimethylation of lysine 27 on histone H3 (*H3K27me3*) frequently keep critical developmental genes switched off. New CRISPR-based epigenetic techniques enable researchers to gently remove these molecular keys by targeting chromatin-modifying enzymes to specific genes, such as *LEC* or *BBM*. This increases regeneration ability without permanently altering the DNA sequence, which is especially useful in monocots and woody plants with higher chromatin barriers (Ikeuchi et al. 2016). Synthetic regeneration engineering represents a shift from trial-and-error methods toward targeted biological design. Totipotency is no longer viewed as a rare outcome of hormone treatment but as a controllable state within a gene regulatory network. Plant regeneration is becoming more predictable and efficient by using CRISPR-based activation, synthetic promoters, gene circuits and epigenetic editing, opening new possibilities for innovation in plant biotechnology.

7.2 CRISPR-Based Transcriptional Activation Systems

7.2.1 Principles of CRISPRa

CRISPRa is an improved version of CRISPR technology that does not cut DNA but rather turns genes on in a precise and regulated manner. In this method, a catalytically inactive dCas9 is fused to transcriptional activator domains such as *Viral Protein 64* (*VP64*), tripartite transcriptional activation complex combining *VP64, p65 activation domain* and *Rta protein (VPR)*, or *SuperNova Tag (SunTag)*. Instead of creating double-stranded breaks, the dCas9 complex is directed to promoter regions, where it activates the plant's own transcriptional machinery and enhances gene expression (Chavez et al. 2015; Lowder et al. 2018). This technique is particularly useful in plant regeneration research because it allows scientists to activate native developmental genes without introducing any constitutive promoters. Rather than permanently overexpressing a morphogenic gene, CRISPRa enables researchers to gradually improve its natural expression at the right time and place. Using tissue-specific or inducible systems, activation may be spatially or temporally regulated by developing guide RNAs that bind to promoters of totipotency-related genes.

CRISPRa also has the advantage of multiplexing. Multiple guide RNAs can be delivered simultaneously, allowing for the coordinated activation of many genes in a developmental network. For example, co-activating *WUS* and *BBM* has been found to significantly improve somatic embryogenesis and shoot regeneration in resistant crops, thereby avoiding lengthy callus phases (Lowe et al. 2016; Khakhar et al. 2018). This is similar to how plant developmental programs naturally function, using multiple regulatory modules rather than relying on the function of a single gene. Recent improvements such as scaffold RNA systems and epigenome-modifying CRISPRa variants further enhance transcriptional strength and specificity (Souza et al. 2019). These advances suggest that regeneration can increasingly be treated as a programmable genetic circuit rather than a purely hormone-driven phenomenon.

7.2.2 Development Regulators as Synthetic Switches

Plant morphogenic regulators act as master switches, controlling cell fate transitions. Among the most significant are *BBM* and *WUS*. *BBM* increases embryogenic competence by converting somatic cells into embryonic identity, whereas *WUS* creates and maintains stem cell sites inside the meristems. When these genes are synthetically activated simultaneously, they may convert differentiated cells into embryogenic cells, significantly increasing transformation efficiency in crops like maize and sorghum (Lowe et al. 2016). Other key regulators contribute to specific phases of reprogramming. *LEC1* and *LEC2* are essential for embryonic identity formation and storage reserve accumulation, which promote embryogenic development (Long et al. 2022; Peng et al. 2025). *WOUND-INDUCED DEDIFFERENTIATION 1* (*WIND1*) connects stress signaling to cellular reprogramming, mediating dedifferentiation after injury (Ikeuchi et al. 2017). Synthetic activation of these regulators can mimic natural wound- or stress-induced regeneration pathways, but in a more predictable and controllable manner. One of the most attractive features of CRISPRa-based activation is its reversibility. Because promoter sequences are not permanently changed and no double-strand breaks are created, inducible CRISPRa systems can temporarily activate totipotency genes. Once regeneration is completed, gene expression can be turned off. This helps to prevent developmental problems such as abnormal embryo formation, enlargement, and sterility, which are all major adverse consequences of constitutive morphogenic gene overexpression (Fig. 7.1). In the future, combining CRISPRa with inducible promoters (chemical, heat-shock, or light-responsive systems) and tissue-specific expression modules might result in finely controlled regeneration platforms that can be modified to different genotypes. Such programmable systems have the potential to significantly minimize genotype dependence in transformation and accelerate crop development processes.

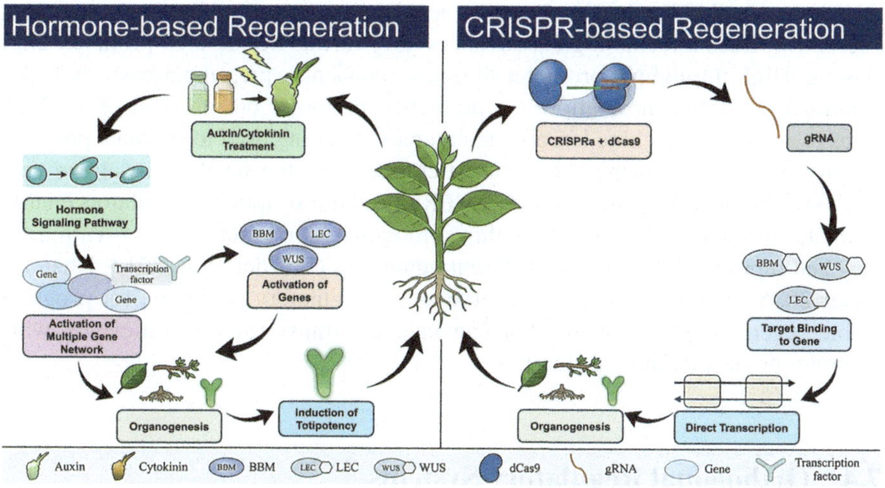

Fig. 7.1 Comparative mechanisms of plant regeneration: Hormone-induced vs. CRISPR-activated pathway

7.3 Synthetic Promoter Engineering

The main objective in plant regeneration engineering is to activate developmental regulators with specific spatial, temporal, and quantitative control. Regeneration efficiency often depends not only on the genes which are activated but also on how accurately their expression is regulated. Therefore, synthetic promoters have become essential tools for establishing predictable and controllable gene expression. Synthetic promoters are often formed by combining a small core promoter (sufficient to draw basal transcription machinery) with engineered cis-regulatory motifs that bind to specific transcription factors. Researchers may build promoters that are tissue-specific, inducible, variable in strength, and have little impact with endogenous gene networks by modifying, stacking, or altering these patterns (Liu and Stewart 2016). In regeneration systems, such design flexibility provides significant advantages. For example, tissue-specific promoters can restrict CRISPRa components or morphogenic regulators to reprogram the explant tissues. Inducible promoters that respond to dexamethasone, estradiol, ethanol, or heat-shock-responsive systems provide temporal control over developmental regulators including *BBM* and *WUS* (Wang et al. 2020). This controlled activation prevents developmental abnormalities that may result from continuous overexpression. Another major advantage of synthetic promoters is modularity. Promoter strength can be calibrated by altering motif number, spacing, or sequence composition. This tunability reduces the pleiotropic effects frequently observed with strong constitutive promoters such as cauliflower mosaic virus 35S promoter (CaMV 35S), which can affect normal development when driving morphogenic genes (Koukara and

Papadopoulou 2023). Rationally constructed promoters provide a more balanced alternative. Recent technological advances have further accelerated promoter engineering. High-throughput promoter libraries, transient expression assays, and computational modeling now allow for predictive promoter design (Cai et al. 2020; Jores et al. 2020). Machine learning techniques are being used to estimate promoter activity based on sequence characteristics, reducing the need for trial-and-error methods. Further, integration with CRISPR-based transcriptional regulators enables synthetic promoters to function within orthogonal gene circuits, improving accuracy and reproducibility across different genotypes (Lowder et al. 2018; Khan and Lister 2025). Synthetic promoters serve as the regulatory backbone for programmable regeneration platforms that can enhance transformation efficiency while maintaining developmental stability.

7.4 Orthogonal Regulatory Systems

One of the biggest challenges in plant synthetic biology is to avoid unintentional interference with the plant's own regulatory networks. Plant genomes are highly interconnected; activating or suppressing one gene can unintentionally influence many others. Orthogonal regulatory systems are designed to solve this problem by functioning independently of the plant's native transcriptional machinery. An orthogonal system uses synthetic transcription factors that recognize DNA sequences that are not found in the plant genome. These artificial binding sites are added to the upstream of target genes, resulting in a closed regulatory module. Because endogenous plant transcription factors cannot bind to these sequences, crosstalk with native pathways is minimized. Technologies such as CRISPR/dCas-based transcriptional regulators and Transcription Activator-Like Effector (TALE)-based activators have been used for this purpose (Lowder et al. 2018; Khan and Lister 2025). In these systems, the programmable DNA-binding domain directs activation to engineered promoter regions, resulting in high specificity. This separation is very important in regeneration engineering. Developmental regulators such as *BBM*, *WUS*, or *LEC* genes are deeply linked with complex hormonal and transcriptional networks. Direct manipulation using endogenous promoters may disturb normal growth patterns, leading to pleiotropic effects or instability. Orthogonal systems allow researchers to insulate synthetic regeneration circuits from native gene networks. As a result, gene activation becomes more predictable, less noisy and easier to fine-tune (Koukara and Papadopoulou 2023). Another major advantage of orthogonality is modularity. Synthetic regulatory modules can be stacked or layered to create logical control systems such as inducible switches, feedback loops, or multi-gene activation cascades without disrupting essential developmental pathways. For instance, a chemically inducible orthogonal transcription factor can activate a synthetic promoter controlling *BBM* expression. Once regeneration is complete, removing the inducer shuts the system down, restoring normal developmental balance. This reversible and insulated control improves both biosafety and developmental stability. Burbano et al. (2024) have also demonstrated the feasibility of building multilayered gene circuits using orthogonal CRISPR-based systems, including programmable activators and repressors that operate simultaneously without interference. Instead of depending solely on hormone gradient and stress response, these developments bring regeneration engineering closer to a circuit-based framework where cell fate transitions may be precisely regulated. Overall, orthogonal regulatory systems provide a critical foundation for building robust, programmable regeneration platforms. By insulating synthetic modules from endogenous gene networks, they enhance reliability, reduce unintended effects and support the development of the next-generation crop transformation technologies.

7.5 Artificial Gene Circuits

7.5.1 Circuit Design Relevant to Regeneration

Traditionally, regeneration has mostly depended on adjusting hormone concentrations and hoping for the desired cells to respond. This approach can be unpredictable and genotype-dependent. Synthetic circuit design offers a more controlled alternative. Instead of simply stimulating cells, it aims to guide them much like programming a system with defined rules and responses. In natural plants, cell fate changes are controlled by complex gene regulatory networks that continuously interpret environmental and internal signals (Brophy et al. 2022). The goal of synthetic circuit architecture is to imitate or optimize these natural systems with greater precision. By designing circuits with defined inputs and outputs, experts can stabilize, redirect, or fine-tune developmental states during regeneration.

7.5.1.1 Toggle Switches: Stabilizing Cell Identity

A toggle switch is a regulatory system that allows a cell to remain in one of two stable states. In regeneration, these two states can be seen as differentiated and totipotent. Positive feedback maintains the cell in an embryogenic state even after the initial signal has vanished until a trigger, such as a chemical inducer or stress signal, activates the switch. The cell will revert only if a deliberate signal is introduced (Moris et al. 2016). This idea fits well with the concept that cell identity behaves like a stable state within a bigger regulatory network. A toggle switch could help maintain embryogenic competence triggered by regulators like *BBM* or *WUS* without requiring continuous hormone treatment (Ikeuchi et al. 2016). By reinforcing the new identity through feedback, regeneration becomes more stable and less random, reducing variability between genotypes.

7.5.1.2 Boolean Logic Gates: Integrating Multiple Signals

Plant cells rarely respond to a single signal in isolation. They interpret combinations of hormones, stress cues, and developmental information before organ formation or embryogenesis. Boolean logic gates recreate this integrative behavior in engineered systems. An AND gate ensures that regeneration happens only when two or more specific signals are present, for example, a wound response and an applied chemical inducer (Zalatan et al. 2015; Koukara and Papadopoulou 2023). This prevents accidental activation in unintended tissues. An OR gate offers flexibility, allowing either environmental stress or a synthetic trigger to start embryogenesis. An NOR gate provides strict control by suppressing regeneration unless inhibitory factors are absent. Building these multi-input circuits in plants is theoretically possible due to developments in synthetic promoters and CRISPR/dCas-based transcriptional regulators (Lowder et al. 2018; Burbano et al. 2024).

7.5.1.3 Pulse Generators: Transient Activation for Precision

One of the major problems in regeneration engineering is that continuous overexpression of morphogenic regulators can cause abnormal growth such as ectopic embryos, excessive meristem formation, or sterility. Studies suggest that only a brief activation of regulators like *BBM* or *WUS* may be sufficient to initiate reprogramming (Lowe et al. 2016). Pulse-generating circuits are designed to provide this short burst of expression. After a defined time window, the system automatically shuts down. Inducible promoters combined with feedback-controlled CRISPR activation systems make this approach possible (Brophy et al. 2022). By limiting activation to a controlled pulse, experts can trigger reprogramming while preserving long-term developmental stability.

7.5.1.4 Feedback Loops: Maintaining Balance

Feedback is central to all living systems. Positive feedback strengthens a newly established state, helping totipotent cells maintain their identity long enough to complete embryogenesis. Negative feedback, on the other hand, ensures that once the regeneration program has progressed, gene activation declines and organized meristem development proceeds normally (Somssich et al. 2016). In natural plant development, stem cell niches are maintained through carefully balanced feedback interactions between transcription factors and signaling pathways. Introducing similar synthetic feedback loops into regeneration circuits improves robustness, reduces noise, and ensures proper timing of developmental transitions. This balance prevents uncontrolled growth while maintaining regeneration efficiency (Del Vecchio et al. 2016). Overall, toggle switches, logic gates, pulse generators, and feedback loops provide a structured framework for programmable plant regeneration. Instead of relying only on hormonal signals, these circuit designs allow regeneration to be controlled through defined inputs and regulated responses.

7.5.2 Circuit Modules in Regeneration Design

A programmable regeneration platform can be designed using a modular framework in which different regulatory components perform specific functions. Similar to engineered biological circuits, each module contributes to sensing signals, processing information, activating developmental programs, and terminating the response at the appropriate stage. This modular architecture increases precision and predictability in regeneration engineering. The input module functions as the sensing unit of the system. It detects signals such as wounding, environmental stress, hormonal changes, or externally applied chemical inducers that initiate regeneration. In synthetic biology approaches, inducible or stress-responsive promoters can be used to respond specifically to these signals. For example, wound-inducible promoters or chemically inducible systems such as dexamethasone- or estradiol-based promoters precisely control the timing and location of regeneration activation (Wang et al. 2020). By ensuring that regeneration begins only under defined conditions, the input module improves spatial and temporal control. After signal detection, the processing module integrates these inputs and determines whether regeneration should proceed. This module often incorporates regulatory logic similar to Boolean logic gates (AND, OR, NOR). Such systems allow cells to interpret multiple signals simultaneously before committing to a developmental response. For example, regeneration may occur only when both a wound signal and an external inducer are present, resembling an AND gate configuration. Advances in CRISPR/dCas-mediated transcriptional regulation have made it possible to construct such multi-input regulatory circuits in plants (Lowder et al. 2018; Burbano et al. 2024). This signal integration helps prevent unintended activation and increases the reliability of regeneration programs. The effector module carries out the regeneration program by activating key morphogenic regulators that drive cellular reprogramming. *BBM*, *WUS*, and *LEC* genes are significant transcription factors that promote embryogenic competence and meristem formation. Activation of these regulators can be achieved through synthetic transcription factors, CRISPR activation systems or engineered promoters that drive controlled gene expression (Lowe et al. 2016; Ikeuchi et al. 2016). The shutdown module ensures that regeneration does not continue indefinitely after the desired developmental transition has occurred. Prolonged expression of morphogenic regulators can cause developmental abnormalities such as ectopic embryos or abnormal meristem growth. As a result, synthetic circuits often incorporate negative feedback loops, inducible repression systems, or regulatory elements that terminate gene expression once a threshold is reached. This controlled shutdown enables the plant to move from embryogenic growth to normal development while maintaining genetic stability (Koukara and Papadopoulou 2023). Overall, combining input, processing, effector, and shutdown modules provides a systematic framework for programmable regeneration. Such circuit-based approaches represent an emerging strategy in plant synthetic biology, enabling precise control over developmental pathways and improving transformation efficiency across a wide range of plant species.

7.6 Synthetic Epigenetic Reprogramming

Plant cells must undergo significant epigenetic alterations in order to achieve totipotency. Gene activity in differentiated cells is closely regulated by mechanisms such as DNA methylation, histone modification, and chromatin compaction. Many developmental genes remain dormant due to epigenetic modifications. These limiting patterns must be reset during regeneration or somatic embryogenesis in order to reactivate important developmental regulators (Fehér 2019; Ikeuchi et al. 2016). During regeneration, several coordinated changes occur at the epigenetic level. Specifically, DNA methylation patterns are altered, accompanied by a decrease in repressive histone marks and a concurrent increase in activating marks at loci associated with embryogenesis. For instance, the repressive mark *H3K27me3* is often reduced, while activating marks such as *H3K4me3* and histone acetylation become more prominent. These changes open the chromatin structure and allow transcription factors to access regulatory regions and initiate gene expression needed for cellular reprogramming (He et al. 2012; Ikeuchi et al. 2019).

Recent advancements in synthetic biology have created new techniques for directly modifying these epigenetic barriers. CRISPR-based epigenetic editors use a catalytically inactive dCas9 fused with chromatin-modifying enzymes. These systems, guided by specific RNA sequences, can target particular gene promoters and alter their chromatin state without causing DNA breaks or permanent mutations (Hilton et al. 2015; Lowder et al. 2018). Different enzymes can be linked to the dCas9 protein to achieve specific epigenetic effects. For instance, DNA demethylases can remove methyl groups from cytosine residues, restoring gene activity. Histone acetyltransferases can loosen chromatin by adding acetyl groups to histones, while histone methyltransferases can introduce activating methylation marks that support gene transcription. Through such targeted modifications, developmental genes can be reactivated without altering the DNA sequence itself. This approach is very useful for improving regeneration in recalcitrant crops. In many monocot species, chromatin around morphogenic regulators remains tightly packed, limiting the expression of genes necessary for regeneration. To overcome these obstacles and improve embryogenic competence, recent strategies have focused on directing epigenetic editors to the promoters of key regulatory genes, such as *WUS*, *BBM*, and *LEC* (Lowe et al. 2016; Ikeuchi et al. 2016). Combining epigenetic editing with CRISPRa is a potential future path. CRISPRa stimulates transcription directly, whereas epigenetic editors modify chromatin to make the DNA more accessible. Together, these approaches can activate developmental programs and improve regeneration efficiency (Koukara and Papadopoulou 2023; Khan and Lister 2025).

7.7 Future Direction of Plant Regeneration

Advances in plant synthetic biology are gradually transforming regeneration from an empirical, hormone-based process into a programmable and controllable system. Technologies such as CRISPRa, synthetic promoters, artificial gene circuits, and epigenetic editing tools provide unequaled control over gene expression and developmental pathways. When incorporated, these tools can form complex regeneration platforms capable of precisely activating morphogenic regulators and guiding cell fate transitions. Such programmable systems have the potential to overcome many limitations associated with conventional tissue culture–based regeneration methods (Brophy et al. 2022; Khan and Lister 2025).

7.7.1 Genotype Independent Transformation

A major challenge in plant biotechnology is the strong dependence of transformation and regeneration efficiency on genotype. Many crop species, especially cereals and woody plants, show large variation in regeneration potential among cultivars. Typically, only a few elite lines are susceptible to efficient transformation, limiting the application of genome editing and transgenic technologies in breeding programs. Synthetic activation of key morphogenic regulators such as *BBM* and *WUS* offers a promising strategy to address this limitation. Lowe et al. (2016) have demonstrated that controlled expression of these regulators can significantly enhance transformation efficiency and enable regeneration in recalcitrant genotypes (Lowe et al. 2016). By incorporating inducible regulatory elements and CRISPR-based activation systems, it may become possible to standardize regeneration protocols across diverse cultivars. Such genotype-independent transformation methods would significantly increase the use of genome editing tools in crop improvement.

7.7.2 In Planta *Regeneration*

Traditional plant transformation is primarily based on tissue culture procedures such as callus induction, regeneration, and acclimatization. These procedures are labor-intensive, time-consuming, and often associated with somaclonal variation. As a result, scientists are increasingly exploring *in planta* regeneration strategies, where new meristems or organs are induced directly within intact plant tissues. Synthetic activation of developmental regulators in situ has demonstrated potential for inducing de novo meristem formation without the need for prolonged callus phases. For example, transient expressions of morphogenic regulators can directly induce new shoots or embryos from plant tissues, allowing altered cells to regenerate into whole plants more rapidly (Maher et al. 2020). Such approaches could significantly simplify transformation operations and make genome editing more accessible for a wider range of crops.

7.7.3 Integration with Speed Breeding

Another interesting path is to combine programmable regeneration systems with speed breeding, a technique that accelerates plant growth cycles through optimized environmental conditions such as extended photoperiods and controlled temperature conditions. Speed breeding allows multiple generations of crops to be produced within a single year (Watson et al. 2018). When integrated with programmable regeneration and genome editing tools, this approach could dramatically shorten crop improvement timelines. In the future, morphogenic regulators could induce embryogenesis, CRISPR systems could introduce targeted genetic modifications, and regenerated plants could rapidly advance through generations using speed breeding techniques. This integrated technique could enable rapid trait stacking and accelerated development of improved crop varieties. The combination of genome editing, synthetic promoters, epigenetic editing, and gene circuit design is paving the way for next-generation plant regeneration platforms. By enabling precise control over developmental pathways, these technologies could reduce genotype dependency, simplify transformation procedures, and accelerate breeding cycles.

7.8 Ethical, Regulatory, and Commercial Consideration

The development of programmable regeneration technologies also involves important ethical, regulatory, and commercial considerations. Engineering plant totipotency raises ethical concerns, as synthetic activation of embryogenic pathways has the potential to alter natural reproductive patterns, gene flow dynamics, and ecological interactions if such traits are unintentionally expressed outside controlled environments (Ikeuchi et al. 2016; Fehér 2019). Regulatory frameworks play a crucial role in governing the application of these technologies. The classification of regeneration engineering approaches depends on whether foreign DNA is permanently incorporated into the plant genome. Technologies based on transient CRISPRa or epigenetic editing that do not introduce stable transgenes may be regulated under gene-editing guidelines rather than conventional transgenic regulations. However, synthetic gene circuits and engineered regulatory elements still require biosafety evaluation to determine possible off-target effects, unintended gene activation, and environmental persistence (Brophy et al. 2022; Khan and Lister 2025). In India, regulatory policies are evolving to distinguish between site-directed nuclease 1 (SDN1) and site-directed nuclease 2 (SDN2) genome edits and transgenic modifications, which significantly influences approval and commercialization pathways for genome-edited crops. From a commercial perspective, regeneration engineering has great potential for agricultural biotechnology. Improved regeneration efficiency can reduce crop transformation time and cost, while universal regeneration platforms may allow genome editing across several cultivars. At the same time, intellectual property rights associated with morphogenic regulators such as *WUS* and *BBM*, along with CRISPR technologies, are likely to shape technology access and licensing models (Lowe et al. 2016; Maher et al. 2020).

7.9 Conclusion

Plant regeneration remains a fundamental component of modern plant biotechnology, enabling genetic transformation, genome editing, and rapid crop improvement. However, conventional regeneration systems are often inefficient, genotype-dependent, and difficult to standardize across plant species. The emergence of regeneration engineering, supported by advances in synthetic biology, offers new strategies to overcome these long-standing challenges. Technologies such as CRISPR-based gene activation, synthetic promoters, orthogonal regulatory systems, artificial gene circuits, and epigenetic editing tools are providing remarkable control over plant developmental pathways. By precisely regulating morphogenic regulators and developmental networks, these approaches help guide cell fate transitions in a predictable and controlled manner. Such innovations are gradually shifting regeneration from an empirical practice to a more rational and design-based process. Importantly, the integration of these technologies may enable the development of universal and programmable regeneration systems. These technologies have the potential to reduce genotype dependence, simplify transformation operations, and enable *in planta* regeneration strategies that do not require significant tissue culture stages. When combined with rapid breeding techniques and modern genome editing tools, regeneration engineering has the potential to significantly accelerate crop improvement programs. The complexity of plant developmental networks, potential off-target effects, regulatory uncertainties, and biosafety considerations must be carefully addressed before widespread application. Continued interdisciplinary research combining plant biology, synthetic biology and computational modeling will be essential to refine these technologies. Collectively, regeneration engineering is a significant step toward next-generation plant biotechnology, providing novel strategies for increasing transformation efficiency, expanding the variety of transformable crops, and supporting sustainable agricultural development in the future.

References

Brophy JA, Magallon KJ, Duan L, Zhong V, Ramachandran P, Kniazev K, Dinneny JR (2022) Synthetic genetic circuits as a means of reprogramming plant roots. Science 377(6607):747–751

Burbano DA, Kiattisewee C, Karanjia AV, Cardiff RA, Faulkner ID, Sugianto W, Carothers JM (2024) CRISPR tools for engineering prokaryotic systems: recent advances and new applications. Annu Rev Chem Biomol Eng 15(1):389–430

Cai YM, Kallam K, Tidd H, Gendarini G, Salzman A, Patron NJ (2020) Rational design of minimal synthetic promoters for plants. Nucleic Acids Res 48(21):11845–11856

Chavez A, Scheiman J, Vora S, Pruitt BW, Tuttle M, PR Iyer E, Lin S, Kiani S, Guzman CD, Wiegand DJ, Ter-Ovanesyan D, Church GM (2015) Highly efficient Cas9-mediated transcriptional programming. Nat Methods 12(4):326–328

Debernardi JM, Tricoli DM, Ercoli MF, Hayta S, Ronald P, Palatnik JF, Dubcovsky J (2020) A GRF–GIF chimeric protein improves the regeneration efficiency of transgenic plants. Nat Biotechnol 38(11):1274–1279

Del Vecchio D, Dy AJ, Qian Y (2016) Control theory meets synthetic biology. J R Soc Interface 13(120):20160380

Fehér A (2019) Callus, dedifferentiation, totipotency, somatic embryogenesis: what these terms mean in the era of molecular plant biology? Front Plant Sci 10:536

He C, Chen X, Huang H, Xu L (2012) Reprogramming of H3K27me3 is critical for acquisition of pluripotency from cultured Arabidopsis tissues. PLoS Genet 8(8):e1002911

Hilton IB, D'ippolito AM, Vockley CM, Thakore PI, Crawford GE, Reddy TE, Gersbach CA (2015) Epigenome editing by a CRISPR-Cas9-based acetyltransferase activates genes from promoters and enhancers. Nat Biotechnol 33(5):510–517

Ikeuchi M, Ogawa Y, Iwase A, Sugimoto K (2016) Plant regeneration: cellular origins and molecular mechanisms. Development 143(9):1442–1451

Ikeuchi M, Iwase A, Rymen B, Lambolez A, Kojima M, Takebayashi Y, Heyman J, Watanabe S, Seo M, De Veylder L, Sakakibara H, Sugimoto K (2017) Wounding triggers callus formation via dynamic hormonal and transcriptional changes. Plant Physiol 175(3):1158–1174

Ikeuchi M, Favero DS, Sakamoto Y, Iwase A, Coleman D, Rymen B, Sugimoto K (2019) Molecular mechanisms of plant regeneration. Annu Rev Plant Biol 70(1):377–406

Jores T, Tonnies J, Dorrity MW, Cuperus JT, Fields S, Queitsch C (2020) Identification of plant enhancers and their constituent elements by STARR-seq in tobacco leaves. Plant Cell 32(7):2120–2131

Khakhar A, Leydon AR, Lemmex AC, Klavins E, Nemhauser JL (2018) Synthetic hormone-responsive transcription factors can monitor and re-program plant development. elife 7:e34702

Khan A, Lister R (2025) Synthetic gene circuits in plants: recent advances and challenges. Quant Plant Biol 6:e6

Koukara J, Papadopoulou KK (2023) Advances in plant synthetic biology approaches to control expression of gene circuits. Biochem Biophys Res Commun 654:55–61

Liu W, Stewart CN Jr (2016) Plant synthetic promoters and transcription factors. Curr Opin Biotechnol 37:36–44

Long Y, Yang Y, Pan G, Shen Y (2022) New insights into tissue culture plant-regeneration mechanisms. Front Plant Sci 13:926752

Lowder LG, Zhou J, Zhang Y, Malzahn A, Zhong Z, Hsieh TF, Voytas DF, Zhang Y, Qi Y (2018) Robust transcriptional activation in plants using multiplexed CRISPR-Act2. 0 and mTALE-act systems. Mol Plant 11(2):245–256

Lowe K, Wu E, Wang N, Hoerster G, Hastings C, Cho MJ, Scelonge C, Lenderts B, Chamberlin M, Cushatt J, Wang L, Gordon-Kamm W (2016) Morphogenic regulators baby boom and Wuschel improve monocot transformation. Plant Cell 28(9):1998–2015

Maher MF, Nasti RA, Vollbrecht M, Starker CG, Clark MD, Voytas DF (2020) Plant gene editing through de novo induction of meristems. Nat Biotechnol 38(1):84–89

Moris N, Pina C, Arias AM (2016) Transition states and cell fate decisions in epigenetic landscapes. Nat Rev Genet 17(11):693–703

Peng J, Zhang Q, Tang LP, Xu BJ, Laux T, Zhang XS, Su YH (2025) LEC2 induces somatic cell reprogramming through epigenetic activation of plant cell totipotency regulators. Nat Commun 16(1):4185

Somssich M, Khan GA, Persson S (2016) Cell wall heterogeneity in root development of Arabidopsis. Front Plant Sci 7:1242

Souza GM, Trewavas T, Daloso DDM (eds) (2019) Systems plant physiology: an integrated view of plants life. Prog Biophys Mol Biol 146:1–2

Wang X, Ye L, Lyu M, Ursache R, Löytynoja A, Mähönen AP (2020) An inducible genome editing system for plants. Nat Plants 6(7):766–772

Watson A, Ghosh S, Williams MJ, Cuddy WS, Simmonds J, Rey MD, Asyraf Md Hatta M, Hinchliffe A, Steed A, Reynolds D, Adamski NM, Hickey LT (2018) Speed breeding is a powerful tool to accelerate crop research and breeding. Nat Plants 4(1):23–29

Zalatan JG, Lee ME, Almeida R, Gilbert LA, Whitehead EH, La Russa M, Tsai JC, Weissman JS, Dueber JE, Qi LS, Lim WA (2015) Engineering complex synthetic transcriptional programs with CRISPR RNA scaffolds. Cell 160(1):339–350

The manufacturer's authorised representative in the EU is Springer Nature Customer Service Centre GmbH, Europaplatz 3, 69115 Heidelberg, Germany. If you have any concerns regarding our products, please contact ProductSafety@springernature.com

Printed and bound by CPI Group (UK) Ltd, Croydon, CR0 4YY
13/10/2026
02230757-0002